# The Changing Nature *of* WORK

## IMPLICATIONS FOR OCCUPATIONAL ANALYSIS

Committee on Techniques for the Enhancement of Human
Performance: Occupational Analysis

Commission on Behavioral and Social Sciences
and Education
National Research Council

NATIONAL ACADEMY PRESS
Washington, D.C.

NATIONAL ACADEMY PRESS • 2101 Constitution Avenue, N.W. • Washington, D.C. 20418

NOTICE: The project that is the subject of this report was approved by the Governing Board of the National Research Council, whose members are drawn from the councils of the National Academy of Sciences, the National Academy of Engineering, and the Institute of Medicine. The members of the committee responsible for the report were chosen for their special competences and with regard for appropriate balance.

This study was supported by Contract No. DASW01-96-C-0051 between the National Academy of Sciences and the U.S. Department of the Army. Any opinions, findings, conclusions, or recommendations expressed in this publication are those of the author(s) and do not necessarily reflect the view of the organizations or agencies that provided support for this project.

### Library of Congress Cataloging-in-Publication Data

The changing nature of work : implications for occupational analysis.
    p. cm.
    "Committee on Techniques for the Enhancement of Human Performance, Commission on Behavioral and Social Sciences and Education, National Research Council."
    Includes bibliographical references and index.
    ISBN 0-309-06525-9
    1. Work. 2. Labor market. 3. Diversity in the workplace. 4. Occupations—Forecasting. 5. Industrial sociology. I. National Academy Press (U.S.) II. National Research Council (U.S.). Committee on Techniques for the Enhancement of Human Performance.
    HD4901 .C428 1999
    331.25—dc21
                                        99-6682

Additional copies of this report are available from:
National Academy Press
2101 Constitution Avenue, N.W.
Lock Box 285
Washington, D.C. 20055
Call 800-624-6242 or 202-334-3313 (in the Washington Metropolitan Area).

This report is also available on line at **http://www.nap.edu**

Printed in the United States of America

The National Academy of Sciences is a private, nonprofit, self-perpetuating society of distinguished scholars engaged in scientific and engineering research, dedicated to the furtherance of science and technology and to their use for the general welfare. Upon the authority of the charter granted to it by the Congress in 1863, the Academy has a mandate that requires it to advise the federal government on scientific and technical matters. Dr. Bruce M. Alberts is president of the National Academy of Sciences.

The National Academy of Engineering was established in 1964, under the charter of the National Academy of Sciences, as a parallel organization of outstanding engineers. It is autonomous in its administration and in the selection of its members, sharing with the National Academy of Sciences the responsibility for advising the federal government. The National Academy of Engineering also sponsors engineering programs aimed at meeting national needs, encourages education and research, and recognizes the superior achievements of engineers. Dr. William A. Wulf is president of the National Academy of Engineering.

The Institute of Medicine was established in 1970 by the National Academy of Sciences to secure the services of eminent members of appropriate professions in the examination of policy matters pertaining to the health of the public. The Institute acts under the responsibility given to the National Academy of Sciences by its congressional charter to be an adviser to the federal government and, upon its own initiative, to identify issues of medical care, research, and education. Dr. Kenneth I. Shine is president of the Institute of Medicine.

The National Research Council was organized by the National Academy of Sciences in 1916 to associate the broad community of science and technology with the Academy's purposes of furthering knowledge and advising the federal government. Functioning in accordance with general policies determined by the Academy, the Council has become the principal operating agency of both the National Academy of Sciences and the National Academy of Engineering in providing services to the government, the public, and the scientific and engineering communities. The Council is administered jointly by both Academies and the Institute of Medicine. Dr. Bruce M. Alberts and Dr. William A. Wulf are chairman and vice chairman, respectively, of the National Research Council.

# Contents

# Preface

In 1985, the Army Research Institute (ARI) asked the National Academy of Sciences to explore the utility and effectiveness of various techniques to enhance human performance. The Academy, through the National Research Council (NRC), established the Committee on Techniques for the Enhancement of Human Performance. The committee, then composed primarily of psychologists, first examined and then evaluated commercial and proprietary techniques then being considered by the Army; later the committee broadened its inquiry to study a variety of related issues, including team learning, simulation training, and skill practice. In 1995, the focus shifted to the organizational context of individual and group performance.

The committee's current topic and the subject of this book are the changing nature of work and the implications for occupational analysis. The charge to the committee from the Army Research Institute was (1) to review and analyze the research on the environmental forces, organizational factors, and the content of work; (2) to identify key issues in the changing context and content of work that affect the design of occupations in the civilian and military sectors; (3) to evaluate the changes in tools for analyzing the nature of the work environment and developing occupational classification systems that are responsive to current and future needs of the workplace; and (4) to assess the application of methods and tools developed in the civilian sector to occupational classification and analysis in the Army. The current composition of the committee includes experts in the areas of sociology, economics, management, occupational analysis, and industrial and organizational psychology and training.

This book is intended to provide decision makers in both public and private organizations, as well as in both the civilian and military sectors, with guidance on how to assess and respond to contemporary debates about changes in work. Our audience extends far beyond the boundaries of social scientists and human resource specialists who have a professional interest in understanding changes in work and the adequacy of occupational analysis systems for charting and managing the changes. In particular, we hope that decision makers whose choices influence the nature of work—who include senior executives, line mangers, military officers, and designers of technology—will find valuable information in this volume.

We extend thanks to Michael Drillings, Chief, Research and Advanced Concepts, at the U.S. Army Research Institute, our sponsor, for his interest in the topic of this report and our work. He has supported the committee's efforts and provided assistance in obtaining information. A number of individuals throughout the U.S. Army helped arrange field trips and provided the committee with special briefings and demonstrations: Dr. Robert Bauer, Directorate of Training and Doctrine Development; CSM Garvey, Commandant, NCO Academy; Walter Gunning, U.S. Army Personnel Command; MG George H. Harmeyer, Commanding General, U.S. Army Armor Center and Ft. Knox; and MG Alfonso E. Lenhardt, Commanding General, USAREC.

Many people have contributed to this study by drafting background material, and those who played the most direct role in developing the manuscript deserve special mention: Grace McLaughlin, University of California, Irvine; Siobhan O'Mahoney, Stanford University; Yuri Suárez and Dillon Soares, Michigan State University; Eva Skuratowicz, University of California, Davis; and Michael Strausser and Danielle Van Jaarsveld, Cornell University.

In the course of preparing this report, each member of the committee took an active role in drafting chapters, leading discussions, and reading and commenting on successive drafts. We are deeply indebted to them, for their broad scholarship, their insights, and their cooperative spirit. Their commitment to our collective work was real and their efforts are immensely appreciated. Truly, our report is the product of an intellectual team effort. Specifically, David Neumark assumed major responsibility for sections on workforce demographics, Peter Cappelli for structure of the material on changes in the organizational contexts of work, Norman Peterson for the chapter on occupational analysis, and Mark Eitelberg for sections on work and demographics in the Army. Kenneth Spenner contributed significant portions of the sections on demographics and on occupational analysis. Robert Vance performed analyses on the Gantz Wiley survey data and contributed significantly to the chapter on occupational analysis. Key materials were contributed by Rosemary Batt on service work, Nicole Biggart on interactive-emotion work, Anne Howard on training, Arne Kalleberg on changing employment relationships and the meaning of work, Paul Osterman on managerial work, and Lyman Porter on teamwork. LTG Theodore Stroup contributed significantly to the identification of Army implications of civilian work trends and identification of means by which the Army uses its occupational classification system to manage its personnel.

This report has been reviewed in draft form by individuals chosen for their diverse perspectives and technical expertise, in accordance with procedures approved by the NRC's Report Review Committee. The purpose of this independent review is to provide candid and critical comments that will assist the institution in making the published report as sound as possible and to ensure that the report meets institutional standards for objectivity, evidence, and responsiveness to the study charge. The review comments and draft manuscript remain confidential to protect the integrity of the deliberative process.

We wish to thank the following individuals for their participation in the review of this report: Eileen Appelbaum, Economic Policy Institute, Washington, D.C.; David J. Armor, Institute of Public Policy, George Mason University; John Campbell, Department of Psychology, University of Minnesota; Wayne Cascio, Graduate School of Business, University of Colorado; Paula

England, Department of Sociology, University of Pennsylvania; Richard Jeanneret, Jeanneret & Associates, Houston, Texas; Frank Landy, SHL Landy Jacobs, Boulder, Colorado; Paul Sackett, Department of Psychology, University of Minnesota; and Charles Tilly, Departments of Sociology and Political Science, Columbia University. Although the individuals listed above provided many constructive comments and suggestions, it must be emphasized that responsibility for the final content of this report rests solely with the authoring committee and the institution.

Staff of the National Research Council made important contributions to our work in many ways. Three staff members contributed to the initial stages of the committee's work: Daniel Druckman, study director, Mary Ann Statman, research associate, and Sharon Vandivere, senior project assistant. We extend particular thanks to Susan McCutchen, the committee's senior project assistant, who was indispensable in organizing meetings, arranging travel, compiling agenda materials, coordinating the sharing of information among committee members, and managing the preparation of this report. We are also indebted to James McGee, who provided help whenever it was needed and who made significant contributions throughout the report. We also thank Christine McShane, who edited and significantly improved the report.

Thomas A. Kochan, *Co-Chair*
Stephen R. Barley, *Co-Chair*
Anne S. Mavor, *Study Director*
Committee on Techniques for
the Enhancement of
Human Performance:
Occupational Analysis

# The
# Changing
# Nature
# *of* WORK

# Executive Summary

U nderstanding work and how it is changing is essential to all decision makers and advisors who educate and prepare people for the workplace, counsel them about career choices, assign them to jobs, and shape the organizational and institutional contexts in which people work. One source of important data used by these decision makers comes from occupational analysis and classification systems. Traditionally, occupational analysis provided a picture of the structure of work, the characteristics of jobs, and the skills and knowledge generally associated with effective job performance. All such systems reflect the historical and cultural milieu in which they were devised and thus, when work changes substantially, the conceptions of work they characterize are inevitably outdated. Unfortunately, a variety of problems arises for human resources professionals engaged in job classification, counseling, selection, and training when job descriptions do not comport with the changes occurring in the workplace.

Although there is a great deal of debate about the changing nature of work in today's society, there is also much agreement that changes have occurred and will continue to occur. A key concern is what these changes imply for the development of a responsive set of occupational analysis tools, methods, and classifications. This book examines the evidence for what is chang-

*1*

ing, the factors that are influencing the changes, and the characteristics of an occupational analysis system that provides human resources personnel with the opportunity to make effective decisions in the face of change.

The impetus for the work of the committee was the recognition by the Army that advances in approaches to occupational analysis and classification might assist personnel managers in the Army to be more responsive to changes in the military work environment and thus more effective in the selection, training, and deployment of enlisted personnel and officers. Since many of the forces influencing work in the civilian sector are also influencing work in the military—such as advances in technology, the introduction of a more diverse workforce, and the extension of missions into new areas—it seemed useful to examine trends and developments in the civilian sector as a source of guidance. A key concern was to assess the applicability of the latest developments in occupational analysis and classification technology to building an effective, flexible, forward-looking system for the Army, one that could be used to monitor changes in the nature of work and assist in the design of new jobs.

This volume presents a framework that integrates sources of occupational change with a generally accepted conceptualization of how occupational analysis reflects and affects the nature of work. It suggests that changes in the content and structure of work are shaped by environmental trends that include changes in markets, technologies, and workforce demographics. Although these forces sometimes affect the content and structure of work directly, they also have indirect effects insofar as they create pressures for organizational restructuring and change in employment relationships. The content and structure of work, in turn, both dictate the kinds of knowledge, skills, and abilities that employees are likely to require and also affect important outcomes, such as the quantity, quality, and efficiency of work; the performance of organizations; and the psychological, social, and economic rewards people achieve through work.

Work structures and occupations are also shaped by the tools and systems of occupational analysis that are used to describe and measure the structure and content of work and to design jobs. To be useful, these systems must be updated frequently enough to keep up with the pace of change in how work is being structured. Moreover, given advances in the technologies available for displaying and communicating how alternative tasks might be combined, the committee believes that it may be possible for these systems to become more forward-looking and serve as analytic aids to decision makers. Without an occupational analysis system that can detect changes in the structure and content of work, it will be difficult to know whether work is changing until the change is complete, at which point attempts to fashion the change are moot.

The committee's major conclusions regarding debates about jobs and work, implications for occupational analysis, the Army, future research directions, and policy follow. A detailed discussion of these conclusions, including a summary of key findings, is presented in Chapter 7.

## CONCLUSIONS

Four themes emerged from the committee's analysis. Three concern increasing heterogeneity of workers, work, and the workplace, and the fourth focuses on the need for a systematic approach to understanding how work is changing. First, the workforce is becoming more diverse with respect to gender, race, education, and immigrant status; these changes appear to have resulted in greater heterogeneity within traditional occupational categories. Second, the boundaries between who performs which jobs and the employment outcomes and experiences of individuals working in different occupations are becoming more fluid. The evidence suggests that both military and civilian organizations are using a wider variety of workers and skills to accomplish their goals. Third, the range of choices around how work is structured appears to be increasing, and these decisions are interdependent. The fourth and final theme flows from this interdependency. The notion that decision makers' responses to changing markets, demographics, and technologies, the human resource policies and systems employed in organizations, and the work structures and outcomes they produce for organizations are interrelated leads to the need for an integrated, systematic approach to understanding how the context of work is changing and the implications of these changes.

## Broader Debates About Jobs and Work

### Singular Trends Versus Constrained Choices

The evidence cautions against making definitive statements about unidirectional trends in the nature of work. Instead, we see increased variance within occupations and multiple options for shaping jobs and for grouping them into occupations. These findings suggest that the future of jobs will not be determined solely by the forces of technology, demographics, or markets but by the interaction of these forces with the strategies, missions, organizational structures, and employment policies that decision makers implement in specific settings. Thus, choice remains important even when options are constrained by external events and when consequences for organizations, individuals, and society are imperfectly predictable.

### The End of Jobs?

Nothing in the data examined by the committee supports the conclusion that all the changes in today's workplace add up to "the end of jobs" in any sense of this term. The conditions and content of work are certainly changing in sometimes dramatic ways, but the vast majority of people in America who want or need to work remain employed. Employment, labor force participation rates, and hours of work have either increased or remained stable in recent years. There is no compelling evidence to suggest that this will change in the future. Moreover, the history of technology repeatedly shows that, even when large numbers of individual workers are driven from particular jobs as a result of a shift in the demand for labor, aggregate demand for workers does not decline because of technical change.

### A Transformation of Work?

Taken alone, none of the changes or trends discussed in this book constitutes anything that could be characterized as a transformation of work. But when combined, as seen in some settings, these changes may lead to new conditions and to possibilities that some might characterize as a transformation. For example, in situations in which markets are uncertain and goals are unclear, work is likely to be more productive if it allows high discretion, flexibility, and the opportunity to work in teams to solve problems, analyze data, and negotiate over courses of action or the meaning of information.

One of our objectives has been to develop a framework that researchers, organizational decision makers, advisers, occupational analysts, and individuals can use as they experiment with designing work, occupational structures, and employment policies. The absence of a clearly articulated framework that includes the full range of forces has limited our ability to assess the combined effects of various changes on individuals, organizations, and society itself. We believe that the social and organizational implications of the combination of changes that we identified need to be examined more fully and systematically by decision makers in both the civilian and military sectors. By explicitly taking into account the full range of factors that shape how work is done, we believe that decision makers have the opportunity to develop more effective alternative work structures that could potentially meet a broad range of needs and interests.

### Implications for Systems of Occupational Analysis and Classification

To adequately track the changing nature of work, occupational analysis and classification systems must take into account the attributes of the persons who perform work, the processes by which they perform it, and the outputs they produce within the dynamic economic, demographic, and technological contexts, and organizational factors that interact with all three. To achieve this objective, occupational analysis and classification systems must widen their traditional scope of attention as well as deepen their level of descriptive detail to capture both the range of relevant factors and the distinctions between jobs and occupations that might otherwise go unnoticed.

Occupational analysis systems must also be sensitive to the greater variance in how work is done within occupations today. Rather than provide a single description of a given job, an adequate system for occupational analysis may need to attend to various alternatives for structuring work in a given job family as well as to the attributes and skill requirements associated with these alternatives.

Shifting from a backward-looking to a forward-looking system that will aid decision makers in designing work structures will also require occupational analysts to consider the human resource and organizational practices needed to support alternative ways of structuring work. By being flexible enough to address future changes in the context and content of work, occupational analysis and classification can contribute not only to the description of work, but also to research that interprets and predicts changes and to the work designs that anticipate those changes.

Over the last several years, the Department of Labor has been developing a system called O*NET™, a prototype for a new approach to collecting, distributing, and analyzing occupational data. In the committee's view, O*NET™ offers several important advances over prior systems in its organization of job description variables and associated data collection instruments, in its electronic databases with job incumbent and occupational analyst ratings, and in the initial technical evaluations. If fully developed and widely used by practitioners who add their own features to the system, we anticipate that it can serve the functions called for here.

- First, O*NET™ is the first available system with planned national scope that brings together the most current category and enumerative systems and the most comprehensive descriptive analytical systems and makes the data readily accessible in electronic format.
- Second, O*NET™ has a theoretically informed and initially validated content model with a more detailed set of job descriptors than other available systems.
- Third, the O*NET™ database can be accessed and used through multiple windows or modes, including entering using job titles or occupations at varying levels of hierarchical detail, but also entering at the level of work descriptors (i.e., knowledge, skills, abilities, other contextual factors). The latter window of

access is extremely important in a world of work that is changing. It allows the analyst or user to build up inductively to the level of job or occupation, in contrast to systems that proceed deductively, starting with a job or occupational category that is anchored in the past and may not be current in its ratings or job descriptive information. O*NET™ could be developed into a decision support tool that allows analysts to compare different models for organizing work, to generate a list of complementary changes needed to support these models, and to project the consequences of these alternatives for the outcomes of central interest to different stakeholders. This feature is perhaps one of the major developments of the O*NET™ prototype.

• Fourth, O*NET™ offers a significant improvement over earlier systems, particularly ones based on the *Dictionary of Occupational Titles*, in the ease of conducting cross-occupational analyses and comparisons.

• Fifth, by utilizing the cross-walks supplied by the National Occupational Information Coordinating Committee, the O*NET™ system allows mapping to other major category and enumerative systems, including military occupational specialties and the Standard Occupational Category system.

Based on these advances, the committee recommends that O*NET™ should continue to be developed as a fully operational system for use in both civilian and military sectors.

## Implications for the Army

The committee's review of the changes in civilian occupations and organizations was organized to provide the Army with a framework for examining its own work structures and occupational analysis systems. The Army is experiencing a number of changes in the context of work that parallel changes experienced in the private economy. We suspect that these developments will also create pressures for change in the structure and content of soldiers' work. These pressures should create opportunities for the military commanders to adjust existing work systems, should they choose to take advantage of them.

**Trends**

The Army's work structure is the basis for selecting, training, organizing, and managing personnel to meet mission requirements. The result of changing mission requirements has been the development of a smaller, more flexible force with a wider range of fighting skills—as well as new skills in negotiation and interpersonal interaction. The increased diversity of Army missions coupled with downsizing has led to the creation of teams composed of individuals from different work cultures with different skills. Some specific trends include:

• The workforce is becoming more diverse with a growing role for women, the increasing age of military personnel, and the more frequent use of units composed of regular Army, reservists, and civilians.

• The development and employment of advanced technology has created a demand for more highly skilled operators and technicians. It is important that the development of technology be integrated with work systems and human resource practices.

• The trend toward joint operations involving hybrid units may require the branches of the military to develop common work structures, or at least structures that can be easily meshed. New missions, particularly those that involve extensive interaction with civilians, will require new tasks; new knowledge, skills, and abilities; and new forms of organizing.

• Downsizing, in combination with advances in technology, has created pressures for soldiers in the lower ranks to share in decision making traditionally embedded in the officer ranks. Attention should be given to the implications of these new roles and for human resource managers.

Army decision makers need to see the design of jobs, work structures, and occupations as tightly linked to their changing missions, technologies, workforce demographics and family structures, and employment practices. The committee therefore recommends that Army decision makers think about the interconnections among these factors and take them into account in structuring work to meet the mission requirements and the needs of those who will be part of the Army of the future.

## Occupational Analysis

The Army's ability to efficiently manage its personnel, in complex and rapidly changing contexts, would be enhanced by an occupational analysis system that efficiently links workforce capabilities with mission planning and provides the structure for recruiting, training, assignment, and promotion of personnel. Such a system would contain all the information needed for such tasks as assembling a special operation in the field or for developing training requirements for a combined military occupational specialty.

Having considered the advantages of O*NET™, the committee sees that it offers promise for meeting the future occupational analysis needs of the Army. AP*NET, an adaptation of O*NET™ proposed for the Army, has several useful features, including linked readiness, occupations, and training databases that allow easy access to descriptions of training courses that teach a particular skill, to lists of soldiers who have skills and abilities relevant to a particular type of mission, and to Army jobs that have similar requirements. Also, AP*NET would be menu-driven with a user-oriented interface that allows access to data at the level of aggregation and specificity that is best suited to each application.

The committee recommends that the Army consider building a prototype of a system whose functional capabilities include those in the AP*NET concept.

## Implications for Research

### Need for Multidisciplinary Studies of Work

In the committee's view, cross-disciplinary collaboration is essential to future progress in the study of work and occupational analysis. However, we do not suggest that all individual studies should abandon their disciplinary focus or traditions. Instead, communication and dialogue across disciplines is needed to inform both the framing of questions and the interpretation of results from multiple disciplines.

### Rethinking Images of Work and Occupations

To gain the full advantage of the opportunities available from new technologies and organizational forms and the changes in the characteristics of the labor force, images of work and the categories used to differentiate among jobs need updating to better reflect: (1) the diversity of the workforce, (2) the dominance of the service economy, (3) the growing role of cognition and analysis, interactions and relationships, and digital technologies in the work people do, and (4) the blurring of the traditional boundaries across which work was divided in the industrial era. The blue-collar-managerial divide in particular no longer captures what people do at work. How to adapt practices, institutions, and public policies that rely on this divide or the other outmoded images are major issues for future study and action.

### Need to Study What Workers Do

Changing the images of work and going beyond abstract arguments about trends in skills requires detailed and rich description and data reported from direct experiences of workers. Thus the sociological and anthropological traditions of observing and participating in real work settings and producing detailed narratives describing the actual experiences of workers need to be encouraged, with the objective of updating perspectives on work. But to be representative, these studies must examine the full array of occupations and workers found in the labor force today. Researchers are especially limited in their ability to describe what managers do at work because it is difficult to measure. Furthermore, sociologists, industrial relations experts, anthropologists, and others continue to focus their efforts on the more easily quantifiable jobs in lower-level occupational groups. It is also important to examine ways of integrating data describing what workers and managers do from other disciplines, such as industrial and organizational psychology and human factors.

### Need for a National Database on Work

Direct observation and in-depth descriptions of what workers do are necessary but not sufficient inputs to update and continue to monitor changes in the aggregate structures of work and the content of jobs. To do this requires a national sample representative of the labor force. This type of data collection is required both to complete the data collection and analysis needed to make O*NET™ operational and to realize its potential and to track systematically the changes in work and their consequences for organizations, individuals, and society.

### Need to Study Occupational Analysis Tools as Aids to Decision Makers

The committee's vision is of a forward-looking occupational analysis system that can be used by decision makers to monitor changes in work, design new jobs, formulate effective human resources policies, and provide timely career counseling. Advances in technology that allow for the consideration of large numbers of variables in a relational database have made it possible to include information not only about jobs and skills, knowledge and abilities, but also about the organizational and environmental forces that influence work. Furthermore, it is now possible to display and combine data to develop what-if scenarios as an aid to job design. In the committee's view, the use of occupational analysis tools to shape work is an extremely important and fruitful area for research and experimentation.

### Implications for Policy

Throughout this study, we note that the laws and institutions governing work and employment largely reflect their industrial-era origins. It goes well beyond the scope of this effort to suggest what changes are needed to update employment laws and institutions to better support work and employment relations today. However, this book may provide a starting point for the analysis of the role of law by presenting data on how work has changed since the basic legal framework governing employment relations was enacted.

# 1

# Introduction

Work and its role in society has become the subject of considerable public commentary and debate in recent years. Some people believe that the world of work is changing so thoroughly and quickly that we should consider ourselves pioneers of a new historical era. Some say that the idea of a job has become antiquated (Bridges, 1994; Arthur and Rousseau, 1996), that job security has become illusory (Osterman, forthcoming), and that, as manufacturing gives way to service work, the workforce will be populated by more unskilled workers (Ritzer, 1998; Levin et al., 1990). Some commentators have even argued that work is disappearing altogether, at least for a significant percentage of the population (Rifkin, 1995; Aronowitz and Cutler, 1998). There are those who assert, more optimistically, that America is becoming a nation of highly skilled technicians, professionals, free agents, and telecommuters (Handy, 1989; Barley, 1996a; Pink, 1998). Still other voices argue that all such claims are overstated, either naively utopian or unnecessarily alarmist. These analysts see changes in the nature of work as more gradual and evolutionary, and that society is experiencing incremental and in many ways expected adaptations to shifts in demography, technology, markets, organizational structures, and employment practices (Farber, 1995).

Our purpose in this book is to clarify the terms of this debate by providing guidance on how to understand changes in work and effectively assess their implications for the changing structure of occupations in the United States. To this end, we examine existing evidence on how the nature of work and the composition of the workforce are changing in both the civilian and the military sectors and review existing systems of occupational analysis and classification. We then delineate the implications of our findings about the changing nature of work for developing tools and approaches to occupational analysis and classification. Our intent is to sort through what is changing and what is not in order to provide an interpretive framework that will aid both organizational decision makers and members of the workforce as they contemplate decisions about career selection and progression. We suggest that by better understanding the full array of forces affecting work and further developing the occupational analysis tools to take account of these changes, decision makers may be better equipped to shape work, occupations, and organizations for generations to come. Indeed, a better understanding of how the forces driving change interact with labor market policies and institutions and with organizational and individual decisions can give decision makers greater control over the future of work and its consequences for individuals, organizations, and society.

## FRAMEWORK FOR ANALYZING THE CHANGING NATURE OF WORK

To better describe and track the nature of work, and possibly to gain greater control over how people work, it is first necessary to understand and consider the full range of forces that shape work and how these forces are changing. Figure 1.1 lays out the framework we use to analyze these forces and their effects on work.

When people speak of "the nature of work," they usually refer to one or more of four tightly related aspects of a society's primary mode of production. The first is what people do for a living, the *occupations* or primary lines of work that characterize a society at a particular point in time. Second is the *content of work* or how people do what they do: the techniques, technologies,

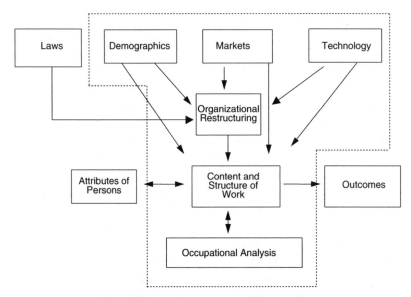

FIGURE 1.1   Framework for conceptualizing the changing nature of work and occupational analysis.

and skills they employ.   Third are the *organizational, social, and institutional contexts* in which work takes place.   Finally, the nature of work encompasses *the way work affects and relates to other aspects of daily life—the standards of living it produces for workers and their families, its relationship to community life, its effects on one's self-esteem and social status, etc.*

Figure 1.1 seeks to understand these different aspects of work by first examining the key external factors affecting work and how it is changing.   For example, one of the most widely recognized forces shaping work is *technology*.   Indeed, one of the reasons there is so much debate today over how work is changing is that some are persuaded that the world is in the midst of another era of transformation in the technological infrastructure similar to those that triggered the first industrial revolution in the late 1700s and the early 1800s and the second industrial revolution at the end of the 19th century.   For example, by the beginning of the 20th century, interchangeable parts, electric power, the electric motor, dedicated machine tools, the internal combustion engine, the telephone, and a number of new office technologies (including the typewriter and the vertical filing system) had occasioned a chain of events that directly and indirectly transformed the nature of work in all four senses described above (Coombs, 1984; Hounshell, 1984; Yates, 1993). New occupations such as automobile mechanics and electricians were born, and existing occupations such as clerical work became increasingly mechanized and differentiated into such subspecialties as typist and filing clerk. The context of work was also irrevocably transformed by the advent of large corporations and urbanization that, in turn, created the milieu for unionization.

Several investigators argue that current changes in the nature of work driven by the multiple uses of digital technologies (digitization) are symptomatic of a third industrial revolution (Bell, 1973; Dertouzos and Moses, 1979; Nora and Minc, 1981; Perrole, 1986; Negroponte, 1995; Block, 1990; Stewart, 1997). They observe that the technology of microelectronics, robotics, and computer-integrated manufacturing, the advent of artificial intelligence, experimentation in electronic data exchange, and the explosion of digital telecommunications evidenced by the unprecedented growth of the Internet and the World Wide Web have brought the world to the verge of a transformation similar to the second industrial revolution.

Three other external forces are frequently identified as contributors to the changing nature of work. These are the demography of the workforce, the globalization of markets, and the laws and regulations governing work and employment relations. The changing *demography* of the workforce includes the growing presence of women, especially young mothers, in the labor market; increasing racial and ethnic diversity, including a declining majority of white workers; an increasing number of dual-career families; increasing levels of educational attainment; and the aging of the workforce. These demographic trends are well documented; not only do they increase the heterogeneity of the working population, but they also create pressures for expanding existing lines of work and for creating new ones to address the needs of a labor force that were previously handled outside the paid economy, through the family and the community.

*Globalizing product markets* creates greater and more uncertain competitive pressures, larger labor markets, and the tendency toward specialization in an international division of labor. In order to regain competitiveness, American and European firms have embarked on a quest for increased flexibility. Key components of this flexibility include lean production and quality management, downsizing, the outsourcing of business services, the use of contract labor, and the growing acceptability of strategic alliances, even among competitors.

A full treatment of the effects of the laws and regulations governing work and employment lies well beyond the scope of this analysis. Although we recognize that legal structures and their enforcement play important roles in the workplace, we diagram these forces outside of the dotted line in Figure 1.1 to signify that we will not deal with them specifically in our analysis. Indeed, we do so because it is our judgment that changes in law should be informed by analysis of changes in how work is being performed. Thus, this report may be a useful input to some future analysis of the implications of changes at work for public policies regulating work and employment.

Having said this, we make two points about how legislation and workplace regulations intersect with our analysis. First, the basic framework for labor and employment law in place today originated from the New Deal legislation of the 1930s. Thus, much of it still reflects the images of industrial work and organizational structures that dominated at that time. That is, the dominant image of work embedded in most of these laws and regulations carried over from that era—the unemployment compensation system, the National Labor Relations Act, the Fair Labor Standards Act, etc.—is that of either a blue-collar production worker or white-collar managerial employee working for a large firm in a long-term employment relationship with a spouse at home attending to personal and family affairs. Thus, it may be problematic that the workforce, family structures and household work hours, and the organization of work have changed considerably since the 1930s while the basic structure of labor and employment law has remained essentially the same.

Second, although the structure of labor and employment laws has not changed, they have expanded considerably in number, scope, range of issues regulated, and complexity. In addition to those cited above, the workplace is now regulated by laws and regulations governing equal employment opportunity, safety and health, pension security, family and medical leave, and many other aspects of the employment relationship. One study estimated that the number of regulatory programs administered by the U.S. Department of Labor expanded from 43 in 1960 to 134 by 1974 and has continued to grow since then (Dunlop, 1976). Moreover, the growth in legislation and regulations has outpaced the ability of agencies and courts to resolve complaints of violations filed under these statutes, with long backlogs of cases reported at agencies such as the Equal Employment Opportunity Commission, the Wage and Hour Division, and the unit responsible for administering the Family and Medical Leave Act, among others (Commission on the Future of Worker Management Relations, 1994). Thus, there are good reasons to believe that the current structures and content of workplace regulations are having two effects at work: (1) they may be trying to force the organization of work into outmoded categories and structures and (2) the overall complexity and difficulty of enforcing these regulations may be producing frustrations on the part of both workers and employers. There is considerable evidence that both of these are true (Commission on the Future of Worker Management Relations, 1994, 1995).

Thus, although beyond the scope of our analysis, these issues may warrant a study of their own at some point in the near future. Our work may serve as a starting point for the analysis of the role of law by providing data on how work has changed since the basic legal framework governing employment relations in the United States was enacted.

These external forces do not uniquely determine the nature of work. Rather, as illustrated in Figure 1.1, we see them as operating through and interacting with the strategies, structures, and processes of the organizations that employ workers. *Organizational restructuring* can also exert an independent effect on work. Indeed, a major theoretical and empirical focus of social scientists who study work and employment relations today is to understand the nature of the interactions between these environmental and organizational forces. This reflects a growing recognition that how people work is influenced in significant ways by the senior executives, managers, and military officers who make choices about how to pursue their organization's strategies and missions. In fact, the successes and failures of these strategies and missions are influenced by the work structures that their decisions create. From this perspective, current experiments with organizational forms, ranging from teamwork to virtual organizations and contingent employment, are as momentous as the rise of corporate and administrative bureaucracies in the late 19th century (Chandler, 1977; Jacoby, 1991; Guest, 1997; Ulrich, 1997).

This interdependence is not well understood, and even when it is recognized in theory, it has not been effectively translated into decision-making processes that join issues of strategy and work design. For example, organizations may address personnel and job design issues after making decisions about competitive strategies or basic missions, choice of technology, and the deployment of other resources. The goal of this report is to provide decision makers with a framework for thinking about how their strategies and decisions affect and reflect work structures and why organizational processes need to be designed to take these interrelationships into account.

One of the main tools used by organizational decision makers to structure work into jobs is *occupational analysis*. Yet, as our review of these systems suggests, their main use to date has been to describe how work has been structured in the past. That is, by their very nature these systems are backward-looking, since they describe jobs at the point in time when data on them were collected. To be useful these systems must therefore be updated frequently enough to keep up with the pace of change in how work is being structured. Moreover, given advances in the technologies available for displaying and communicating how alternative tasks might be combined, we envision the possibility for these systems to become more forward-looking and serve as analytic aids to decision makers. This vision, however, is mostly speculative at this point in time and, as we note in the final chapter, stands as one clear implication and need for future research and experimentation. Thus, occupational analysis systems, if properly updated and fully utilized can support better tracking and assessments of changes in work, perhaps helping to improve decision making in organizations and the counseling of individuals.

Together, these external and organizational forces and the

tools of occupational analysis shape how people work—what we label in Figure 1.1 the *content of work*. The content and structure of jobs, in turn, dictate the kinds of knowledge, skills, and abilities that employees are likely to require and also affect important outcomes, such as the quantity, quality, and efficiency of work; the performance of organizations; and the psychological, social, and economic rewards people achieve through work.

## NATURE OF THE EVIDENCE

Although it is possible to draw conclusions about why the nature of work and occupations are changing, it is difficult to say with certainty what the changes imply. First, data on the way in which work is changing are ambiguous. For instance, one can point simultaneously to studies that suggest that the new world of work will require greater cognitive and interpersonal skills and to other studies that suggest that fewer such skills will be needed than in the past. Frequently, these conflicting images arise because researchers have used different methods or have studied different types of work (Spenner, 1990, 1995), but in some cases researchers have reached conflicting conclusions after studying similar occupations using similar methods (Kuhn, 1989). Similarly, researchers who have studied changing organizational practices differ over whether the use of teams and advanced technologies universally increases (Mankin et al., 1996; Cotton, 1993; Cohen and Bailey, 1997) or decreases (Babson, 1995; Barker, 1993; Graham, 1995) workers' skills and autonomy, or whether the outcomes are contingent on organizational context (Adler, 1993; Appelbaum and Batt, 1994; Berggren, 1992; Klein, 1989).

Second, studies that examine what workers actually do and how they do it are surprisingly rare. Most studies that pass as investigations of the changing nature of work actually examine changing organizational or employment practices; although these certainly affect work, they are not the same as work content or practices. For example, studies of the replacement of full-time employees by temporary employees tell us nothing about what full-time and temporary employees do at work. Although situated studies of work have a venerable history in sociology, the number of sociologists who currently study work activities in detail is quite small, and the number who concern themselves with how work practices are changing is even smaller. Most contemporary sociologists of work have focused on field studies of professional and blue-collar occupations and, as a result, there have been relatively few studies of the most rapidly growing occupational sectors—managers, salespeople, engineers, technicians, and service workers.

Among anthropologists there has been a long-standing taboo against studying work in industrial societies, especially the work of people who are not somehow underprivileged. Hence, anthropological studies of work are even rarer than sociological studies. Finally, despite the fact that industrial psychologists and ergonomists routinely study the details of work, their research often concerns specific skills and abilities at the individual level of analysis and focuses on purposive measurement for such applications as the design of workplaces, employee selection systems, training programs, and appraisal and compensation systems, rather than on attempts to characterize shifts in the nature of work more broadly. As a result, data on work in postindustrial settings are largely, though not exclusively, anecdotal.

The most significant roadblock to assessing whether and to what extent the nature of work is changing arises, however, from the fact that existing systems of occupational analysis and classification are outdated. Occupational analysis refers to the procedures that analysts (usually industrial psychologists) use to characterize the attributes of a specific job and the knowledge, skills, and abilities that are required to perform the job. Occupational classification systems are taxonomies for naming and grouping occupations into ever larger subsets (see Box 1.1 for definitions).

All systems of occupational classification and analysis reflect the historical and cultural milieu in which they were devised. Conk (1978), for example, has shown that, from 1870 to 1940, the Bureau of the Census specifically designed its system of occupational categories to chart the rise of industrialism and its associated problems, especially the effects of immigration on urban demography and the social structure. To the degree that occupational classifications reflect the social and cultural conditions of an era, the conceptions of work they encode will inevitably be-

---

**BOX 1.1**
**Definition of Terms Used in the Report**

**Work:** human activity that is goal directed, purposive, or instrumental and creates value to society. The processes by which humans transform resources into outputs.

**Work structures:** patterns of work activity that are multiply determined by social, economic, political, technological, and cultural processes.

*Work* and *work structures* are manifested in:

**Position:** a single individual performing a particular set of work activities in a particular location.

**Job:** a collection of individual positions having common work activities in a specific employment relationship.

**Occupation:** a collection of individual jobs having similar work activities.

**Occupational structure = occupational category system:** an organization of occupations into categories for some purpose, including labels and connections or relationships between labels.

**Occupational analysis:** the tools and methods used to describe and label work, positions, jobs, and occupations.

**Occupational classification—has two general meanings:** (a) the act of classifying positions, jobs, or occupations into an existing occupational category system and (b) the set of occupational categories in an occupational category system.

---

come outdated. The *Dictionary of Occupational Titles* of the U.S. Department of Labor (Spenner, 1979, 1983; Steinberg, 1990, 1992) makes far more elaborate and accurate distinctions among blue-collar than among white-collar occupations (National Research Council, 1980), although only 25 percent of today's workforce is employed in blue-collar jobs, a percentage that is falling (Silvestri, 1997).

Furthermore, because occupational analysis systems are backward-looking and tend to emphasize skills and attributes that are valued by the culture at the time they are developed, they not only fail to assess attributes of work that may eventually prove to be more important, but they also make attributions that in retrospect seem unwarranted, if not biased. That is, they reflect reality of the past that may not accurately represent the current or future nature of work. For example, Steinberg (1992) has shown that current systems for classifying occupations, such as the *Dictionary of Occupational Titles* and the Hay System, not only use managerial work as a baseline for assessing the complexity of other jobs, but they also tend to rate occupations traditionally filled by men (e.g., zookeeper) as more highly skilled than occupations traditionally filled by women (e.g., child care provider).

Unfortunately, outdated systems of occupational analysis and classification hamper more than the ability to assess the changing nature of work; they also create serious practical problems. As noted earlier, they are used by educators, employment counselors, military planners, policy makers, and parents.

## MULTIDISCIPLINARY APPROACH

Given the uncertain and multiple directions in which the nature of work may be changing and the fact that research on the topic is both scarce and diverse, determining whether and how work is changing and then recommending an approach for mapping an occupational structure carries risks. Particularly important is the failure to consider a sufficient range of perspectives and types of evidence. As a result, we have attempted to incorporate the contributions of a variety of disciplines.

Research on work and the workplace is found in many disciplines, including economics, labor relations, sociology, organization studies, anthropology, industrial and organizational psychology, and ergonomics. The methodological and substantive differences among these disciplines are useful for developing a broad understanding of how and why the nature of work may be changing. As changes occur, they must be examined and understood at three levels of analysis: (1) work and the individual practitioner, (2) organizations and other institutions in which the work is embedded, and (3) the economy and society as a whole. The different disciplines typically have more to say about one level of analysis than the others. In general, anthropology, industrial and organizational psychology, and ergonomics focus on the content of work, whereas economics, industrial relations, and organizational studies concentrate on its context. Sociologists tend to conduct studies in both areas.

Anthropologists (and ethnographically oriented sociologists) are primarily concerned with describing the practices associated with specific types of work and the meaning that those practices have for the individuals who perform them. Most anthropologists argue that work and its meaning are culturally and socially constituted and, for this reason, are likely to vary across contexts. Ergonomists share with anthropologists an overriding concern for the specifics of work practice, but ignore meaning and study instead the physical and cognitive interactions that occur between humans and machines. For anthropologists, work is a social and interpretive activity; for ergonomists, it is a biophysical or an information-processing activity. Industrial and organizational psychologists also study individuals and the jobs they perform. However, unlike either anthropologists or ergonomists, they are relatively unconcerned with the unique aspects of machines or tasks and instead seek to identify and measure individual abilities and task requirements using constructs that generalize across situations. More than those in any other discipline, industrial and organizational psychologists have developed extensive and rigorous methodologies for analyzing jobs for purposes of specifying the skill, knowledge, and ability requirements associated with selection, performance management, and training program development.

The most relevant features of work for an economist are the wages that employers pay, the hours that workers spend, the skills workers bring to the workplace, and the incentives created by employers. Broadly speaking, issues pertaining to work, such as changes in skill requirements, the conditions of work, and the organization of the workplace, can be treated in terms of their implications for wages, skill levels, and the structure of incentives. For instance, if the changing nature of work demands a more skilled workforce, then those who have more skill should garner more pay and returns to skill should increase over time, as workers respond by acquiring more skills. By imposing a market framework on work and work structures, economics provides both a starting point for the analysis of work and an important criterion for evaluating outcomes: Does a work structure increase or decrease the efficiencies and social benefits that flow from the allocation of scarce resources?

Sociologists and industrial relations researchers (as well as institutional economists) concern themselves primarily with the institutions in which work is embedded. These range from the structure of internal labor markets and human resource systems to the laws that govern the employment contract. Thus, like organizational theorists, sociologists and industrial relations researchers are more likely to study the context in which work occurs than its specific content. Sociologists have historically had more to say about occupations than any other discipline and have done the most research on how work and occupations contribute to social stratification. When sociologists speak of the occupational structure, they tend to refer to the division of labor in society; when industrial psychologists and industrial relations researchers speak of occupational structures, they frequently refer to internal labor markets or to the patterns that link jobs across organizations and thus create occupational, regional, or industry wage structures and labor markets.

The various disciplines differ in their approach to work in several other substantive ways as well. Industrial relations and sociology pay more attention to the historical context of work than any other discipline—with the exception, of course, of history itself. Thus, these disciplines have more to say about employment relations institutions and policies because they bring issues of interests, power, and social values directly into their models and analyses of work. Anthropologists and some sociologists tend to view skill and knowledge as properties of communities. Industrial and organizational psychologists, ergonomists, and economists treat skill and knowledge as properties of individuals. Because of its ties to engineering, ergonomics deals most directly with technology's influence on work, but its purview tends to be highly situation-specific. After ergonomics, sociology and anthropology have probably shown the most interest in technology as a material cause. Economists view work as rationally motivated, whereas sociologists and anthropologists are more likely to view work as a normative or cultural activity.

Given the committee's view that a multidisciplinary approach is crucial and that the phenomena of interest are multifaceted, analyses have been guided by the following principles. First, we value and have drawn broadly on different methods and types of data, such as ethnographies, organization-specific case studies, surveys, census data, and other sources of information as a basis for suggesting emerging developments and potential trends. Furthermore, for purposes of illustration, we have relied on examples from field research and well-known events at specific organizations. When data are conflicting or ambiguous, the interpretation is based on our collective judgment. Our hope is that these judgments will encourage further research and debate.

Second, we have refrained from concluding that, because specific developments are occurring today, they will necessarily endure or spread across the economy. We have exercised caution about concluding that a given change represents a fundamental transformation in work structures, and we urge readers to do the same. In our view, fundamental transformation requires widespread breaks from past patterns on multiple, interrelated fronts. At the moment, it is difficult to make the connections necessary to argue how various developments are interrelated or whether they will have long-term consequences.

Third, we do not seek to predict the future. Rather, we have attempted to provide a framework that will help individuals and organizational leaders to make more informed choices about issues concerning labor markets, work, and careers. We consider such a stance to be required because one of our primary findings, emphasized throughout this report, is that there is no singular deterministic trend in how work is changing. Instead, there are systematic variations that can be influenced by choices.

Finally, we found it necessary to restrict the scope of our inquiry by limiting our analysis to paid work. We are acutely aware the large numbers of Americans, especially American women, work in the home or contribute to the larger good of society by working in volunteer positions. By ruling home and volunteer work outside the scope of our inquiry, we have no intention of implying that such activities are not crucial to the economy. They surely are. Rather our decision reflects the fact that there is almost no research on how new technologies and other social changes are altering the nature of home and volunteer work.

Nevertheless, based on what is known about the history of work in the home in the early part of the 20th century, we suspect contemporary changes are afoot here as well (Cowen, 1983; Boris, 1994).

## CHARGE TO THE COMMITTEE

The Army Research Institute asked the National Research Council to form a committee to examine issues related to the changing nature of work and the adequacy of occupational analysis systems to classify them. The committee was asked to perform the following tasks:

• Review and analyze the research on the environmental forces, organizational factors, and the content of work that contribute to an understanding of the nature and structure of work.
• Identify key issues in the changing context and content of work that affect the design of occupations.
• Evaluate the changes in the tools for analyzing the nature of the work environment and developing occupational classification systems that are responsive to the current and future needs of the workplace.
• Assess the application of methods and tools developed in the civilian sector to occupational classification and analysis in the Army.

These issues are taken up in the following chapters by working through the structure laid out in Figure 1.1. Because the committee's objective is to address the question of whether and why the nature of work may be changing and what this implies for occupational analysis, we are less concerned either with the personal attributes that changes in work may require or with the way in which the content and structure of work influence organizational or personal outcomes. The dotted line in Figure 1.1 encloses the relations on which this book focuses. In addition, the committee adopted a common set of definitions to lend consistency to the analysis.

## GUIDE TO THE REPORT

The next chapter presents a general overview of the major external factors that can influence both the organization and content of work. Chapter 3 turns to the organizational context of work and reviews what is known about recent trends in the restructuring of organizations and what these changes in organizational context imply for work and work relationships. Chapter 4 synthesizes recent research on changes in the actual content and structure of blue-collar, service, managerial, professional, and technical work. These chapters constitute our analysis of what is known and unknown about the ways in which work seems to be changing.

Chapter 5 introduces the topic of occupational analysis and explores what the findings of the preceding chapters imply for current and future systems of occupational analysis and classification. Chapter 6 recapitulates our analysis of the civilian sector for the special case of the Army. Chapter 7 details the committee's central conclusions and implications for policy and research.

The detailed discussion in the chapters is guided by four themes, three of which concern increasing heterogeneity. First, the workforce is becoming more diverse with respect to gender, race, education, and other demographic characteristics. The most visible of these demographic changes over the past 20 years has been the increasing participation and expanding role of women in both the civilian and the military workforces. The increasing presence of women in the workforce means that there are more families in which both spouses work and more single parents in the workforce. As a result, the needs of workers and their families are changing in ways that make work and family decisions highly interdependent. We currently do not have very good data on what the increasing presence of women means for the nature of work in most occupations.

Second, the boundaries between lines of work are becoming more fluid. Traditional distinctions associated with vertical (hierarchical) divisions of labor, such as manager-worker and exempt-nonexempt, no longer seem useful as they once were for distinguishing between lines of work. Horizontal divisions of labor (the allocation of distinct duties to specific positions or jobs) are being blurred as organizations emphasize teamwork and hold a larger range of employees responsible for communicating and interacting directly with customers, clients, or coworkers inside or outside the organization. Perhaps as a result of these changes, we observe increased within-occupation variability in compensation, employment security, skill requirements, and other outcomes. Nevertheless, this variation suggests that both military and civilian employers are using a wider variety of people and skill mixes to accomplish their missions and goals. Work today often involves teaming people with different skills, backgrounds, and pay structures who have seldom or never worked together before.

Third, the range of choices about how to structure work appears to be increasing. Although market, technological, and demographic forces impinge in systematic ways on how work structures evolve, they are not deterministic. Organizational decision makers, job design specialists, and the tools they use also shape work structures and occupations. We need to better understand how organizational contexts and employment relations influence the technical processes of job design and occupational analysis.

Fourth, the interdependent nature of work structure decisions highlights the need to take a systemic approach to understanding how the context of work is changing and the effects of these changes on outcomes. Technology, work structure, and human resources policies and practices (selection, training, compensation) are all interdependent. A growing body of evidence suggests that bundles of human resources and work practices have bigger effects on performance than individual practices. We therefore need to provide a conceptual map of the broader system of decisions that need to be considered in thinking about work and occupational analysis.

# 2

# The External Contexts of Work

This chapter provides a brief discussion of markets, technology, and labor force demographics and their roles in shaping the organization and content of work. Work structures vary in the pace by which they are exposed to, are affected by, or react to these forces. Occupational analysts and other decision makers who influence how work is structured need to systematically take into account the full range of factors affecting work structures and the consequences of their actions for the full range of stakeholders involved. How decision makers respond to changing markets, technologies, and demographics, the human resource policies and systems employed in organizations, and the work structures and outcomes they produce for organizations, individuals, and society are all interdependent.

## CHANGING MARKETS[1]

### Change in Product Markets

Since the demand for labor is derived from the demand for the products and services it produces, any effort to understand how work is changing must start with how product markets have

---

[1]The material in this section draws heavily on Cappelli et al. (1997:26-39).

been changing. Increased product market competition associated with globalization and deregulation has brought about two types of change: (1) downward pressure on prices and therefore on labor and other production costs and (2) increased pressure to compete in terms of speed, innovation, variety, and customization.

## Increasing Price Competition

The heightened price competition facing U.S. industry is the result of both increased international trade and the deregulation of domestic industries. Between 1980 and 1995, imports as a percentage of U.S. gross domestic product rose sharply from 8 to 14 percent (*Economic Report of the President*, 1995). U.S. manufacturers faced lower-priced, high-volume goods from low-wage countries as well as relatively lower-priced, high-quality goods from high-wage countries such as Japan. Price competition in manufacturing fueled the demand for cheaper services as inputs and, as a result, many service providers no longer enjoy protected or local markets. Changes in technology and deregulation (in such industries as financial services, transportation, utilities, and telecommunications), accelerated domestic competition among service providers, and the mobility of information technologies coupled with international deregulation in services led to higher international competition in services (Office of Technology Assessment, 1987; McKinsey Global Institute, 1992). Deregulation in service industries has led to an influx of new entrants that have lower cost structures because they: (1) have no sunk costs in outdated technologies; (2) pay lower wages than those negotiated and enforced through collective bargaining in the oligopolistic structures of the regulated industries; and (3) utilize work systems and employment contracts that are more flexible and that in some cases rely more on nonstandard employment arrangements that shift risks associated with market uncertainty from the firm to the workforce (e.g., Belzer, 1994; Keefe and Batt, 1997; Lipsky and Donn, 1987). Deregulation has also increased wage inequality by shifting employment to the nonunion sector, in which wage inequality is greater (DiNardo et al., 1996; Fortin and Lemieux, 1997).

U.S. firms responded to price competition by downsizing, flattening hierarchies, and restructuring organizations and processes. In one survey, for example, three-fourths of the corporate respondents cited increased competitive pressures as the key factor motivating restructuring of their operations (Wyatt Company, 1993). Increases in investments of foreign companies in the United States also grew significantly during this time period (growing from $83 billion in 1980 to $406 billion by 1990) and put additional pressure on production and work systems to compete with ones designed (and sometimes managed) by international competitors.

These changes in markets interacted with—and in industries such as trucking and telecommunications accelerated (e.g., Belzer, 1994; Batt and Keefe, 1999)—the decline in unionization. In industries especially hit by price competition (such as auto supply, steel, tires, apparel, electrical machinery, and trucking), union strategies to "take wages out of competition" by maintaining common wage standards (Kochan et al., 1986) were substantially curtailed. Jobs that in the past paid high premiums could be supplied more cheaply in other countries or in domestic enterprises that pay competitive market rates. As a result, unionized employment fell as these jobs moved to lower-cost environments and organizations. This is particularly true of semiskilled blue-collar work in both manufacturing and services. Thus, the combination of increases in international and domestic competition is one major cause of the restructuring experienced in American industry in recent years. It is also a contributing factor to the increase in inequality in the wage structure that is discussed later in this chapter. Those with the most scarce skills and capabilities realized increasing returns to these attributes, and those with fewer and more easily replaced skills were affected most by pressures to hold down wages and labor costs.

## Product Innovation, Variety, Customization, and Speed-to-Market

Along with increased price competition, markets have changed in ways that require increased capacity and speed in developing and introducing new and more varied products. Product cycle times, for example, have declined significantly in recent years (Fine, 1998), and batch production has risen. U.S. firms have responded by experimenting with a wide variety of new forms of work organization (e.g., Appelbaum and Batt, 1994; Cappelli et al., 1997). For example, to reduce product development time and enhance innovation, firms have introduced various types of cross-functional teams. To improve manufacturing quality and flexibility and reduce time-to-market, they have reduced job classifications, increased job rotation and multitasking, and used shop floor teams and employee participation in problem solving and statistical process control. They have also gained flexibility by focusing on core competencies and efficient supply chain management. Although there is little direct evidence of the effects of decreased product cycle time on job content, the implications are that employees must adapt by continuously learning new skills and new product knowledge in order to produce and service new products. In sum, increased change and variability in product content is associated with new forms of work organization as well as new and more rapidly changing skill requirements of jobs. These changes have profound implications for our occupational classification systems and undermine the extent to which they map the reality of work.

## Changing Financial Markets

Researchers have paid less attention to how financial markets influence work structures than to the effects of product markets. Yet capital markets have always been recognized as having a major influence on the organizational forms that evolve in industries and societies (Chandler, 1977; Roe, 1994; Aoki, 1988). For example, the rise of large, integrated corporations in the late 1800s was made possible by the pooling of large sums of capital and gave rise to the dominant role that finance plays in American firms (Fligstein, 1990; Lazonick, 1992). Similarly, in the 1960s, the growth of large-scale conglomerates in the United States reflected the growing view that firms could diversify their risks by operating in different product markets subject to varied exposures to market fluctuations. In the current period, two changes in financial markets appear to be affecting the content of work: (1) the increased focus on shareholder interests and (2) the increased volatility in international capital flows.

### Shareholder Interests

Developments over the past two decades have begun to focus attention on the role of financial markets and institutions in organizational restructuring and disciplining management to focus more narrowly and intensively on the interests of shareholders and other capital market agents. In the 1980s, the rise in the market for corporate control (Lazonick, 1992; Porter, 1992), the rise of shareholder activism (Useem, 1996), and the growth of agency theory within economics were all contributing factors (Appelbaum and Berg, 1996). For example, the 1980s wave of hostile takeovers, mergers and acquisitions, and leveraged buyouts created pressures on American companies to refocus their resources on their "core competencies" and to sell or close business units deemed nonessential to the company's main product line or service activity. Deregulation of financial markets, the growth of mutual funds, and the increased leverage of institutional investors put pressure on top management and led to new activism among members of corporate boards of directors—leading to the replacement of chief executive officers in a number of large and highly visible companies such as IBM, Eastman Kodak, and General Motors. Corporate boards have also tied an increasingly larger proportion of executive pay to shareholder value. The net result of these pressures was to induce top executives to become more responsive to shareholder and investor concerns (Useem, 1996).

In some cases, firms took preemptive measures and downsized in anticipation of future problems and in anticipation of a favorable stock market reaction, rather than treat downsizing as a strategy of last resort taken only in the face of an immediate crisis (Osterman, forthcoming). The evidence suggests that downsizing alone did not produce favorable reactions from the stock market, but downsizing combined with other restructuring moves did have at least a temporary positive effect on stock price (Worrell et al., 1991; Cascio et al., 1997). Although the direct links between these pressures and work content have not been well researched, it is apparent that these changes in financial markets provided additional incentives for major organizational downsizing and restructuring, which, in turn, profoundly affect the organization of work and content of jobs—by collapsing multiple jobs into single jobs to reduce the number of workers, by dividing jobs in new ways to facilitate outsourcing, or by reorganizing work in new ways to improve productivity and competitiveness.

## Global Capital Flows

In addition to financial market changes that link shareholder interests more closely to managerial decision making, the last two decades have witnessed a dramatic rise in the volume and volatility of global capital flows (Burtless, 1995). The relationship between these changes and managerial decision making with respect to production strategies and work organization also has not been well researched, but is important to recognize. The logical implications are that any given productive enterprise faces a more uncertain future, lacking certainty about what level of organizational performance is sufficient to attract and sustain the commitment of capital to investment in the enterprise. We can only speculate about how this heightened uncertainty influences managerial decision making with respect to the organization of work, adoption of new technology, demand for skills, and deployment of labor. In the case of the information services industry, market and technological uncertainty appears to lead companies to hedge their bets by merging and consolidating, on one hand, and investing in competing technologies, on the other (Keefe and Batt, 1997). But the responses to uncertainty are likely to vary significantly across industries and contexts. At a minimum, however, rising environmental uncertainty is likely to be associated with greater variety in experimentation with organizational forms and greater instability in the content of work, suggesting greater challenges for the occupational classification system to accurately mirror the reality of work.

We explore the combined effects of these financial and product market changes on the organization and content of work in more detail in Chapters 3 and 4. For now it is sufficient to note that they serve as one set of critical drivers of changes in employment relationships and changes in the way work has been carried out in recent years.

## CHANGING TECHNOLOGY

Historically, advances in technology have had profound effects on the workplace and how work is conducted. In essence, technology and work are integrally related (Baba, 1995): work is the processes by which humans transform resources into outputs (Applebaum, 1992), technologies are the means by which the transforming is done (Perrow, 1967). Technology, therefore, shapes not only what people can do, but how they do it. Typically, technological change has three effects on work and occupational structures. It creates new occupations and reduces or eliminates some existing occupations; it increases the skills required on some jobs and decreases the skills of others; and it changes the skills required in ways that are not captured by "up or down-skilling" effects. Because changes in technology occur with some regularity in most work settings, the reformatting of work by new technology is commonplace. Nevertheless, most technological change entails an incremental modification to an existing technological base. The effect of incremental changes can be dramatic, however, as many organizations that rely on computers discover when they switch to a new operating system (DOS, UNIX, Windows) or programming language (COBOL, C++). Incremental technological changes may alter the parameters of specific jobs, seal the fate of particular individuals, and even create or destroy entire occupational communities. But incremental technological changes are unlikely to trigger broad shifts in an occupational structure because they build on, and hence leave unchallenged, existing technological regimes. Broadscale occupational shifts usually require a change in the technical infrastructure.

### Digitization: Change in Infrastructure

The technologies of the second industrial revolution (interchangeable parts, electrical power, the electric motor, dedicated machine tools, the internal combustion engine, the telephone, the typewriter, and vertical filing systems) slowly made obsolete those occupations that were predicated on the previous technological regime, a system of production based largely on hand tools and animate or natural power sources augmented by steam engines and railroads. The question now being debated by historians of technology is whether the advances in digital technologies are ushering in a new era—a third industrial revolution.

With the proliferation of microelectronics, the spread of robotics and computer-integrated manufacturing, the advent of artificial intelligence, experimentation in electronic data exchange, the explosion of digital telecommunications technology, and the unprecedented growth of the Internet and the World Wide Web has come the premonition that the world again stands on the verge of a profound transformation (Bell, 1973; Dertouzos and Moses, 1979; Nora and Minc, 1981; Perrole, 1986; Negroponte, 1995). At the core of this change is digitization, which refers to the conversion of physical phenomena and meaningful symbols like words and numbers into binary (or digital) electronic signals and the use of those signals to control machines and create or manipulate information. The engineering community is currently in the process of fusing the servo-mechanisms characteristic of the second industrial revolution with microelectronics to create a technological base that melds the mechanical and the digital. The Japanese refer to this hybrid technology as "mechatronics." Digitization is a fundamental change in a technological paradigm that is conceptually on a par with electrification (Nye, 1990; Hughes, 1983).

## Digitization and Skills

Over the last two decades, debate and research on the implications of digital technologies for the nature of work has centered on three hypotheses: *deskilling, upgrading,* and *polarization.* The *deskilling* hypothesis predicts that widespread use of digital technologies will result in less skilled, more routine work. Contemporary deskilling theory stems from Harry Braverman's (1974) influential book, *Labor and Monopoly Capitalism.* According to Braverman, by encoding production plans in computer programs, digital technologies enable management to transfer conceptual tasks to programmers and engineers and relegate operators to doing little more than data input or monitoring a machine's actions.

In sharp contrast, the *upgrading* hypothesis predicts that digital technologies will, on balance, create new jobs and transform existing work in ways that demand greater levels of skill. The upgrading thesis rests on two lines of reasoning. The first, rooted in the studies of work in continuous processing industries (Blauner, 1964; Gallie, 1978), argues that highly digitized work environments may eliminate the need for older skills, but simultaneously demand new skills that many jobs did not previously require: for example, responsibilities for monitoring, visualizing, and intervening in an entire production process, for responding quickly and decisively in the case of emergencies, and for interacting with a broad array of people in other functional roles (Adler, 1992; Hirschhorn, 1984; Zuboff, 1989; Kern and Schumman, 1992). The second line of argument is based on the contention that working through a symbolic interface is, more often than not, a substantively complex activity that requires people to have technical skills, to conceptualize transformation processes abstractly, and to analyze, interpret, and act on abstractions instead of, or in addition to, sensory data (Zuboff, 1989; Barley, 1990).

The *polarization* or *mixed change* hypothesis claims that the shift to digital technology pushes in both directions. Some occupations may be deskilled, others may be upgraded, and still others may experience both forces simultaneously depending on a variety of contextual factors (Barley, 1988; Spenner, 1995; Jones, 1982). Versions of this hypothesis that speak of polarization (Gallie, 1994) generally envision a bifurcation of the occupational structure along lines of skill: an increase in both high-skilled and low-skilled work and a gradual elimination of work that falls in between. Advocates of the mixed change hypothesis tend to see little net change in overall levels of skill, because forces for deskilling and upgrading cancel each other when aggregated.

The most convincing data, to date, comes from the Social Change and Economic Life Initiative (SCELI) sponsored by Great Britain's Economic and Social Research Council (Penn et al., 1994). In 1986, SCELI researchers surveyed over 6,000 individuals randomly selected from the electoral registries of six cities in Great Britain. The cities were selected to include those that had experienced both higher and lower than average levels of unemployment. Respondents were asked a wide range of questions concerning their employment experience, including whether they felt the skills they used on the job had increased, decreased, or stayed the same over the preceding five years. Overall, 52 percent of respondents felt that their skills had increased, whereas only 9 percent reported that their skills had decreased. And 60 percent reported that levels of responsibility in their job had grown, whereas only 7 percent reported a decrease (Gallie, 1994). The pattern of results was robust across occupational categories and industries and was similar for people who did and did not change jobs, but evidence of upgrading was least pronounced among low-skilled manual workers in service industries.

Furthermore, the SCELI data clearly show that upgrading was associated with the use of "automated or computerized equipment" (Gallie, 1994:63). The data show that 39 percent of the respondents reported using digital technologies. Of these, 67 percent reported an increase in skill and 74 percent reported an increase in responsibility. Among those who did not use digital technologies, the percentage reporting an increase in skill and responsibility were 39 and 49 percent, respectively. These results were also robust across occupational categories.

The SCELI data on the impact of digital technologies are consistent with attempts to estimate financial returns from computer use. Analyzing data from 1984 and 1989 Current Population Survey, Krueger (1993) found that using a computer in one's job led to a 10-15 percent premium in wages after controlling for obvious covariates, such as years of education, job tenure, industry, and occupation. DiNardo and Pischke (1997) found returns to computer use of similar magnitude using German data. However, they also found that use of other office technologies, including pencils and paper, also increased wages after controlling for computer use. DiNardo and Pischke therefore argue that economic returns to computer use potentially reflect an unmeasured phenomenon, since it is hard to believe that the ability to use pencil and paper is in short supply. Gallie reaches much the same conclusion regarding the use of computers in the SCELI data: "It is unlikely that these trends reflect a deterministic impact of new technology . . . the general association between change and higher skill levels is likely to reflect factors such as the prevailing nature of managerial views about effective ways of enhancing employee motivation and the bargaining power of employee work groups" (1994:65) However, neither study provides evidence of a plausible alternative explanation.

After reviewing the research on changing levels of skill conducted up to the early 1990s, Spenner (1995) concluded that aggregate studies of skill, especially those that focused on changes in the occupational composition of the workforce, were more optimistic about upgrading than was the literature composed of case studies of technical change in specific occupations and organizations. These conflicting findings and methodological inconsistencies led Spenner (p. 81) to conclude that "much of what we . . . know suggests an uncertain, complicated and contradictory relationship between technological change and skill requirements of work. Technology has substantial effects on the composition and content of work . . . but these effects vary for different dimensions of skill, for different jobs, occupations, industries, and firms and for different technologies."

By focusing debate on upskilling and downskilling, this literature largely misses other important changes in the mix of skills required to take full advantage of emerging digital technologies. Moreover, this debate also fails to do justice to the interactive effects technology and work organization have on skill requirements. We have more to say about this in Chapter 4 on the content of work, especially when we review the evidence of the different approaches to technological change and their effects on performance in the automobile industry (Shimada and MacDuffie, 1987; MacDuffie, 1996).

## CHANGES IN WORKFORCE DEMOGRAPHICS

A third factor that influences changes in the nature of work is the changing composition of the workforce (see Figure 1.1). That is, it is unlikely that changes in the nature of work can be examined in isolation from changes in who works, as the composition of the workforce is likely to influence how work is organized and performed. At the turn of the 20th century, the population of the United States numbered about 76 million people, with just under 40 percent of the population in the civilian labor force. By the year 2000, the total population is expected to be in the neighborhood of 275 million people, with a civilian labor force in the range of 150 to 145 million (about 50 percent of the population). This section presents evidence on some of the salient demographic changes in the U.S. population and the U.S. workforce and presents some indirect evidence on how these changes may influence occupational analysis and classification.

There is no claim that demographic changes, in and of themselves, directly lead occupational classifications to become outmoded. More likely, there are several ways in which demographic changes indirectly shape the occupational structure and occupational classification systems and analysis, and hence merit attention. Demographic change shapes who is available to work in the workforce. Significant changes in the types of workers in the workforce, and more importantly in the types of workers performing various jobs, may point to instances in which traditional occupational classifications are likely to break down. For example, classification systems developed in an era when manufacturing and blue-collar jobs traditionally filled by men predominated may represent this sector of the economy in fine detail and afford much less detail to other occupational arenas, such as clerical work (National Research Council, 1980).

Furthermore, demographic changes may directly shape the changing content and contexts of work, and hence indirectly shape occupational analysis and classification systems. For example, there is also a growing body of evidence (Williams and O'Reilly, 1998; Chatman et al., 1998) that demonstrates that demographic diversity (age, race, gender, etc.) affects the social interactions and processes of groups or teams by altering patterns of communications, cohesion, conflict, and decision making, which in turn affect performance. These performance effects can be either positive or negative, depending on how effectively the intervening social processes are managed. Thus, increased diversity may have an effect on work contexts, content, and outcomes through these group or team processes and their management. In this way, increased diversity poses some new challenges to managers.

Another way in which demographic change can indirectly shape the occupational structure and occupational classification systems and analysis is through demands for products and services. For example, the baby boom generation, both as children and as young parents, contributed to the conditions leading to an expansion and differentiation in child care services. Similarly, the aging of the population is among the forces pressing for the expansion and differentiation of health care specializations, health care delivery systems, and related technologies, including expanded institutional options and home health care. One could also include consumer products available to the elderly, ranging from foodstuffs to prescription drugs to entertainment.

In this section, we present major demographic trends in the U.S. population and the workforce (based on data from the Current Population Surveys and the U.S. Bureau of the Census) and data showing the extent to which demographic changes have occurred both across occupations and within occupations (based on original research). It is the changes within occupations that are likely to provide the greatest challenge for existing occupational classification, because the increased heterogeneity of workers within an occupation may be associated with differentiation of tasks within the occupation. For example, the entry of women into police work may result in new tasks and may generate work or skill requirements that reduce the importance of physical strength.

## Trends in the Population and the Workforce

### Age, Fertility, and Family Structure

Over the course of the 20th century, both the U.S. population and the civilian labor force have aged. In 1900, the median age for the white population in the United States was 22.9 years (Gill et al., 1992:76). Estimates for the nonwhite population at the turn of the century are considerably less reliable and generally show a lower median age compared with the white population. By 1997, the median age of the population had risen to 37.3 years for whites, 29.8 years for blacks, and 26.5 years for Hispanics (U.S. Bureau of the Census, 1996b).

One major trend underlying the aging of the population has been the decline in the total fertility rate. A white woman in 1900 could expect an average of 3.56 children by the end of her child-bearing years. Nonwhite fertility rates at the turn of the century were higher but, again, the estimates are subject to greater error. In contrast, by the mid-1990s, a white woman could expect to bear just under two children, whereas black and Hispanic total fertility rates were somewhat higher (Bachu, 1995). By the late 20th century, families were smaller, and there were many more families and households with a single parent. More families and households now are likely to have both spouses working, or the only adult in the family or household working.

### Women and Mothers in the Workforce

Between 1890 and the mid-1900s, the participation rate of women in the labor force increased from 1 in 5 to 3 in 5 (*Monthly Labor Review*, 1997:61; Sweet, 1973). Currently, during the prime working years from ages 25 to 54, about three-quarters of women are working. Figure 2.1 describes the sex composition of the workforce by age in the period 1948-1996. Continuing the longer-term trends, the panel for 25- to 54-year-olds shows the convergence of the representation of men and women in these age ranges in the workforce. Although less pronounced, this same convergence appears for the other three age groups shown in the figure (16 to 24, 55 to 64, and over 65). Thus, at young, old, and prime working ages, women's share of the workforce has increased.

### Racial and Ethnic Changes

At the turn of the century, African Americans were less than 15 percent of the population, and Hispanics and Asians were less than 5 percent (Passel and Edmonston, 1992; U.S. Bureau of the Census, 1992). Furthermore, the major source of immigrants was from Europe—over two-thirds of whom were male. As we approach the end of the century, the origins of immigrants have diversified, with much larger shares from Asia in particular, and

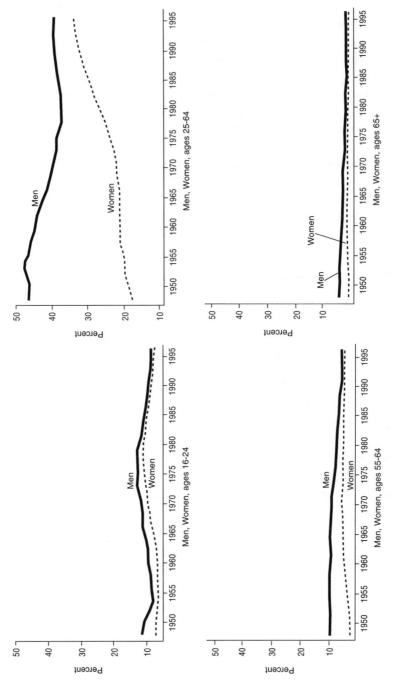

FIGURE 2.1 Percentage distribution of civilian employment by age and sex.

with nearly equal shares of men and women (Gill et al., 1992:320-323). Nonwhites now compose about one-quarter of the U.S. population. Middle Series Projections by the U.S. Census Bureau show the minority share of the U.S. population approaching 35 percent by 2020 and 50 percent by 2050 under a given set of assumptions about trends in fertility, longevity, and immigration (Passel and Edmonston, 1992; U.S. Bureau of the Census, 1992).

## Educational Attainment

The 20th century has seen a substantial increase in the years of formal schooling and educational credentials of both the U.S. population and the labor force. The increases have occurred across all gender and racial groupings, although the average levels of educational attainment differ among whites, blacks, and Hispanics. These differences are even more extreme at higher levels of education. The rising educational attainment of the workforce is apparent in recent data. Between 1970 and 1995, the fraction of the workforce with less than a high school diploma declined sharply. By the mid-1990s, 93 percent of whites, 87 percent of African Americans, and 57 percent of Hispanics had received a high school diploma or equivalency certificate (National Center for Educational Statistics, 1997:92). Also, the fractions of the workforce with some college, particularly with four or more years, rose strongly, especially for women, for whom the fraction was near zero at the beginning of the sample period. These trends in educational attainment, however, do not necessarily point to greater diversity of the workforce. For example, the fraction with a high school education—the modal level of educational attainment—appears relatively stable over this period (at approximately 20 percent).

## Trends in Worklife

### Duration of Work

One way to summarize some of the most important trends is with comparisons of life and worklife expectancy for U.S. men and women for selected years since the turn of the century. Table

TABLE 2.1   Life and Work Expectancy at Birth for Selected Years by Sex
(in years)

| Expectancy | 1900[a] | 1940[b] | 1950[b] |
|---|---|---|---|
| Men |  |  |  |
| Life expectancy | 48.2 | 61.2 | 65.5 |
|    Work expectancy | 32.1 | 38.1 | 41.5 |
|    Nonwork expectancy | 16.1 | 23.1 | 24.0 |
| Women |  |  |  |
| Life expectancy | 50.7 | 65.7 | 71.0 |
|    Work expectancy | 6.3 | 12.1 | 15.1 |
|    Nonwork expectancy | 44.4 | 53.6 | 55.9 |
| Women's worklife as a percentage |  |  |  |
|    of men's worklife | 19.6 | 31.6 | 36.3 |

[a] Data for 1900 are for white persons in death registration states.

[b] Figures adjusted to remove 14- and 15-year-olds from the labor force to be consistent with 1970 (1900 is not comparable).

[c] Figures for 1970 and 1980 are based on the increment-decrement methodology. All other figures reflect ordinary life table methodology. See text for explanation.

2.1 provides the data.[2]   The life expectancies of both men and women increased substantially over the course of the century from about 50 years at the turn of the century to well over 70 years at the end of the century. These increases amount to 20 additional years of life for men and about 30 additional years for women born in 1980.

For men born at the turn of the century, nearly all of their nonwork lives occurred prior to entry into the labor force, with approximately 32 years of worklife and 16 years of nonwork. By 1980, the last year for which methodologically sound estimates exist, men had added about seven years to their worklife expectancy, a small decline from the peak in 1950, after which retire-

---

[2] An earlier version of this argument and the data were presented in Spenner (1988).

| 1960[b] | 1970[c] | 1980[c] | 1990 | 1995 |
|---------|---------|---------|------|------|
| 66.8 | 67.1 | 70.0 | | |
| 41.1 | 37.8 | 38.8 | na | na |
| 25.7 | 29.3 | 31.2 | na | na |
| | | | | |
| 73.1 | 74.8 | 77.6 | | |
| 20.1 | 22.3 | 29.4 | na | na |
| 53.0 | 52.5 | 48.2 | na | na |
| | | | | |
| 48.6 | 59.0 | 75.8 | na | na |

SOURCES: Fullerton, H.N., and Byrne, J.J. 1970. Length of working life for men and women. *Monthly Labor Review*, 1976 (99):31-35 (Table 1) Smith, S.J. 1985. Revised worklife tables reflect 1979-1980 experience. *Monthly Labor Review* (108):23-30 (Table 3). Parts of this table were first published in Spenner (1988).

ment became a widespread transition and men's labor force participation rates edged downward for ages above 55 (Foner and Schwab, 1983; Hayward and Grady, 1990). Thus, much of the increase in life expectancy for men translated into nonwork or part-time work activity after the ages of 55 or 60. Women's patterns of labor market work and nonwork activity show much more substantial change and convergence toward men's patterns. A typical female born in 1900 could expect just over six years of work in her lifetime in the labor force. Over the course of the century, her work expectancy increased dramatically to about 30 years in the labor force by 1980.

As worklife has been increasing, the length of the work week has declined substantially over much of the 20th century, although highly precise time-series data are not available. The best available estimates suggest that the length of the work week declined from about 62 hours per week in 1880 to about 42 hours a week in 1950 (Leontif, 1982). The trend since 1950 is far less certain for several reasons. First, the industrial goods-producing sector, for which the best comparative data exist over the long term, no longer dominates the economy, compared with the rise of service-producing industries in the post-World War II era. Second, some would argue that the changing contexts of work, for example telecommuting, multiple job holding, and work at home, have increased the intensity and duration of work in people's lives, an argument that is considered further in Chapters 3 and 4 (Hochschild, 1997). Hence, the typical length of work week estimates may no longer accurately capture the time that work takes in our lives.

Third, and related to the above point, major debate exists on methodological issues surrounding measurement of trends in hours of work. Conventional survey methodology, as used in the Current Population Survey, shows considerable stability in the average hours worked per week over the past 20 years, with men averaging about 41 hours and women about 35 hours per week in the mid-1990s (Rones et al., 1997). In contrast, some ethnographic studies (Hochschild, 1997) and much of conventional wisdom suggest that people are working longer hours in more places, have less free time, and feel more pressed for time now compared with the past. More comprehensive methodology and data, that is, national time diary studies, show a fair amount of stability in the estimated length of the work week between 1965 and 1985, and perhaps even a slight decline in the length of the work week of men but not of women (Robinson and Bostrom, 1994). The estimates by Robinson and Bostrom (1994) showed a sample of U.S. men in 1985 working an average of 46.4 hours per week, and women working an average of 40.6 hours per week. Schor (1992) criticizes these estimates, suggesting they overestimate women's labor force participation at the start of the period and ignore possible changes in weeks worked per year. Her national data and comparisons show a 9 percent increase in hours worked per year (from 1,786 to 1,949) between 1969 and 1987, for fully employed U.S. workers. In a recent study, Jacobs and Gerson (1998) found that men in professional and managerial careers were more likely to work 50 hours or more per week (34.5 percent) than men in other occupations or women in any occupation. Of women workers in the sample, 17 percent of those in professional and managerial jobs reported working 50 or more hours per week, and approximately 26 percent of those in nonprofessional/managerial occupations reported working less than 30 hours per week. These results were based on self-reports of hours worked in the previous week. Furthermore, Jacobs (1998) suggests that self-reports of work week length correlate highly with independent measures of working time, such as time between departure and return (with commuting time subtracted).

Another relevant dataset is the WorkTrends™ survey, conducted annually since 1984 by Gantz Wiley Research of Minneapolis. This multitopic survey was administered to samples of U.S. households, stratified according to key U.S. census demographics, including age, income, and geography. Surveys were completed by principal wage earners, with the stipulation that respondents were employed full-time at establishments of 100 employees or more. Analyses of the WorkTrends™ database for the years 1985, 1990, and 1996 were prepared for the committee for use in this report. In 1985, there were 5,000 surveys distributed and 2,667 completed and returned (53 percent); in 1990 and 1996, 10,000 surveys were distributed with returns of 4,573 (46 percent) and 6,978 (70 percent), respectively.

## Meaning of Work

Two major concepts are generally used to assess the meaning of work for the individual: (1) *job involvement* or *work commitment* is the extent to which work is a central life interest and (2) *work values* are the extent to which people place importance on various aspects of work. The ability to draw conclusions about trends in these two aspects of the meaning of work is hampered considerably by the paucity of useful data on worker attitudes. One data set that does contain some information on trends in the meaning of work for Americans is the General Social Survey. This survey is a nearly annual, multitopic survey administered to a sample of roughly 1,500 adult, English-speaking American men and women (for an introduction to the General Social Survey, see Davis and Smith, 1992).

*Work Centrality*   One way to assess the extent to which work is a central life interest for workers has been to ask them if they would continue to work if they could live comfortably without working. This question was first asked in a study conducted by Morse and Weiss at the University of Michigan (1955). They found that 80 percent of a national sample of male workers said they would continue to work even if they did not have to do so for financial reasons. They interpreted this finding as demonstrating both that work was a central activity to most American men, and that work meant more to them than simply an economic activity. This basic result has since been replicated a number of times in the United States (for a summary, see O'Brien, 1992).

A form of this question was also asked in the General Social Survey: "If you were to get enough money to live as comfortably as you would like for the rest of your life, would you continue to work or would you stop working?" As Table 2.2 shows, 69 percent of Americans in 1973 said they would continue to work. In

TABLE 2.2   Trends in Work Centrality in the United States, 1973-1996 General Social Survey

| Year | % of Americans Saying They Would Continue to Work |
|------|----------------------------------------------------|
| 1973 | 69.1 |
| 1974 | 64.8 |
| 1976 | 69.0 |
| 1977 | 70.0 |
| 1980 | 76.9 |
| 1982 | 72.3 |
| 1984 | 76.0 |
| 1985 | 69.5 |
| 1987 | 75.4 |
| 1988 | 71.0 |
| 1989 | 72.2 |
| 1990 | 72.7 |
| 1991 | 66.9 |
| 1993 | 69.0 |
| 1994 | 65.8 |
| 1996 | 68.0 |

1996, 68 percent said they would continue to work, which represented virtually no change. In the intervening years, a low of 65 percent (1974) and a high of 77 percent (1980) responded similarly. These results are consistent with the percentages obtained in other surveys of the United States in this time period, which are reported in O'Brien (1992). These data reinforce the Morse and Weiss finding that Americans are highly committed to work as a central activity in their lives.

These findings are supported by the Gantz Wiley Research WorkTrends™ survey. Mean responses by year and occupational grouping are shown in Table 2.3 (the scale is from 1 to 5—1 is the most positive response). Responses to the item, "I like the kind of work I do," although showing minor fluctuations across time and occupational groupings, revealed no consistent trend. Respondents generally reported that they like the work they do (on average, 86 percent agreed or strongly agreed with the item). Professionals and managers tended to be most positive in their responses (approximately 91 percent of professionals and 89 percent of managers agreed or strongly agreed) and laborers were least positive (approximately 69 percent agreed or strongly agreed). Similarly, respondents reported substantial amounts of job satisfaction (on average, 68 percent said they were satisfied or very satisfied with their job), with little change from 1985 to 1996. Again, professionals and managers tended to be most satisfied (approximately 73 percent of each group answered satisfied or very satisfied) and laborers least satisfied (approximately 59 percent answered satisfied or very satisfied).

*Work Values* Trends in work values were assessed in the General Social Survey by asking respondents the following questions: "Would you please look at this card and tell me which one thing on this list you would most prefer in a job?" The card contained the names of five job characteristics and respondents were asked to rank them from 1 (most important) to 4 (fourth most important) (the job characteristic that was not chosen as one of the four most important was coded as 5).

The five job characteristics that were ranked were: high income (income); no danger of being fired (job security); short working hours, with lots of free time (hours); chances for advancement

TABLE 2.3    Trends in Attitudes Toward Work, 1985-1996 Gantz Wiley
Research WorkTrends™ Survey

|  | Years | Total Sample | Clerical | Sales | Service |
|---|---|---|---|---|---|
| I like the kind of work I do.[a] | 1985 | 1.80 | 1.91 | 1.76 | 1.92 |
|  | 1990 | 1.73 | 1.79 | 1.75 | 1.82 |
|  | 1996 | 1.83 | 1.90 | 1.92 | 1.95 |
| Considering everything, how satisfied are you with your job?[b] | 1985 | 2.33 | 2.39 | 2.26 | 2.49 |
|  | 1990 | 2.29 | 2.27 | 2.34 | 2.44 |
|  | 1996 | 2.26 | 2.23 | 2.25 | 2.40 |
| My work gives me a feeling of personal accomplishment.[a] | 1985 | 2.04 | 2.18 | 2.01 | 2.21 |
|  | 1990 | 1.93 | 2.03 | 1.99 | 2.03 |
|  | 1996 | 1.99 | 2.12 | 2.08 | 2.10 |
| My job makes good use of my skills and abilities.[a] | 1985 | 2.34 | 2.43 | 2.33 | 2.51 |
|  | 1990 | 2.20 | 2.29 | 2.26 | 2.41 |
|  | 1996 | 2.11 | 2.21 | 2.30 | 2.33 |
| How do you rate the amount of pay you get on your job?[c] | 1985 | 2.48 | 2.57 | 2.43 | 2.80 |
|  | 1990 | 2.53 | 2.63 | 2.57 | 2.77 |
|  | 1996 | 2.65 | 2.79 | 2.81 | 2.91 |
| How do you rate your total benefits program?[c] | 1985 | 2.05 | 2.02 | 2.04 | 2.00 |
|  | 1990 | 2.19 | 2.04 | 2.26 | 2.32 |
|  | 1996 | 2.27 | 2.15 | 2.41 | 2.42 |
| How do you rate your company in providing job security for people like yourself?[c] | 1985 | 2.13 | 2.14 | 2.24 | 2.09 |
|  | 1990 | 2.28 | 2.27 | 2.45 | 2.20 |
|  | 1996 | 2.70 | 2.69 | 2.71 | 2.71 |
| I am seriously considering leaving my company within the next 12 months.[a] | 1985 | 2.45 | 2.59 | 2.50 | 2.48 |
|  | 1990 | 3.60 | 3.51 | 3.45 | 3.56 |
|  | 1996 | 3.65 | 3.64 | 3.52 | 3.60 |
| In my company, employees are encouraged to participate in making decisions which affect their work.[a] | 1990 | 2.82 | 2.86 | 2.72 | 3.12 |
|  | 1996 | 2.68 | 2.71 | 2.73 | 2.96 |

| Crafts | Operatives | Laborers | Technical | Professionals | Managers and Executives | $R^2$ |
|--------|-----------|----------|-----------|---------------|-------------------------|-------|
| 1.86 | 2.16 | 2.31 | 1.84 | 1.60 | 1.76 | .043 |
| 1.71 | 1.83 | 2.29 | 1.70 | 1.57 | 1.67 | .043 |
| 1.89 | 1.99 | 2.26 | 1.78 | 1.65 | 1.73 | .032 |
| 2.39 | 2.58 | 2.54 | 2.43 | 2.22 | 2.24 | .012 |
| 2.25 | 2.30 | 2.54 | 2.33 | 2.20 | 2.20 | .010 |
| 2.39 | 2.42 | 2.42 | 2.30 | 2.21 | 2.14 | .008 |
| 2.11 | 2.33 | 2.66 | 2.09 | 1.78 | 1.91 | .054 |
| 1.91 | 2.22 | 2.51 | 1.87 | 1.72 | 1.86 | .046 |
| 2.02 | 2.22 | 2.48 | 1.96 | 1.74 | 1.91 | .040 |
| 2.22 | 2.68 | 3.05 | 2.60 | 2.09 | 2.20 | .046 |
| 2.16 | 2.23 | 2.86 | 2.13 | 1.99 | 2.06 | .041 |
| 2.12 | 2.30 | 2.64 | 2.12 | 1.87 | 2.00 | .039 |
| 2.32 | 2.41 | 2.44 | 2.35 | 2.56 | 2.12 | .027 |
| 2.46 | 2.39 | 2.57 | 2.50 | 2.55 | 2.23 | .018 |
| 2.54 | 2.53 | 2.64 | 2.58 | 2.63 | 2.46 | .015 |
| 2.22 | 2.25 | 2.26 | 2.00 | 2.06 | 1.89 | .009 |
| 2.33 | 2.29 | 2.38 | 2.26 | 2.15 | 2.10 | .010 |
| 2.41 | 2.39 | 2.42 | 2.18 | 2.20 | 2.16 | .011 |
| 2.43 | 2.67 | 2.50 | 2.14 | 2.00 | 1.92 | .026 |
| 2.52 | 2.32 | 2.74 | 2.35 | 2.12 | 2.19 | .024 |
| 3.04 | 2.85 | 2.79 | 2.87 | 2.66 | 2.59 | .009 |
| 2.12 | 2.46 | 2.48 | 2.68 | 2.40 | 2.34 | .008 |
| 3.76 | 3.81 | 3.61 | 3.63 | 3.61 | 3.62 | .004 |
| 3.79 | 3.70 | 3.68 | 3.57 | 3.69 | 3.62 | .002 |
| 3.01 | 3.02 | 3.03 | 2.87 | 2.70 | 2.56 | .022 |
| 2.88 | 2.92 | 2.87 | 2.72 | 2.55 | 2.48 | .022 |

*Continued on next page*

TABLE 2.3    Continued

| | Years | Total Sample | Clerical | Sales | Service |
|---|---|---|---|---|---|
| I am given a real opportunity to improve my skills in my company.[a] | 1985 | 2.80 | 2.92 | 2.58 | 3.05 |
| | 1990 | 2.70 | 2.75 | 2.68 | 2.86 |
| | 1996 | 2.68 | 2.70 | 2.62 | 2.85 |
| I am satisfied with the opportunities for training and development that my company provides me.[a] | 1990 | 2.88 | 2.97 | 2.91 | 2.93 |
| | 1996 | 2.82 | 2.82 | 2.79 | 2.94 |
| Considering everything, how would you rate your overall satisfaction in your company at the present time?[b] | 1985 | 2.51 | 2.50 | 2.37 | 2.63 |
| | 1990 | 2.49 | 2.48 | 2.47 | 2.58 |
| | 1996 | 2.51 | 2.46 | 2.44 | 2.66 |
| Senior management gives employees a clear picture of the direction the company is headed.[a] | 1990 | 2.97 | 2.91 | 2.79 | 3.08 |
| | 1996 | 2.79 | 2.68 | 2.54 | 2.90 |
| When my company's senior management says something, you can believe it is true.[a] | 1990 | 3.12 | 3.05 | 3.03 | 3.29 |
| | 1996 | 2.97 | 2.87 | 2.78 | 3.09 |
| Sample sizes (N) | 1985 | 2667 | 482 | 204 | 204 |
| | 1990 | 4573 | 759 | 352 | 366 |
| | 1996 | 6978 | 879 | 370 | 555 |

Notes:  Analyses of the WorkTrends™ data presented in this table were prepared for the committee by Gantz Wiley Research of Minneapolis.  Tabled values are item means, for total samples and occupational subsamples.  Five-point response scales are used for all items, as follows:

[a]1 = Strongly Agree, 2 = Agree, 3 = Neither Agree nor Disagree, 4 = Disagree, 5 = Strongly Disagree.

[b]1 = Very Satisfied, 2 = Satisfied, 3 = Neither Satisfied nor Dissatisfied, 4 = Dissatisfied, 5 = Very Dissatisfied.

| Crafts | Operatives | Laborers | Technical | Professionals | Managers and Executives | $R^2$ |
|---|---|---|---|---|---|---|
| 2.90 | 3.29 | 3.12 | 2.86 | 2.66 | 2.57 | .028 |
| 2.85 | 2.84 | 3.10 | 2.71 | 2.52 | 2.55 | .024 |
| 2.89 | 2.99 | 2.91 | 2.75 | 2.59 | 2.54 | .016 |
| 3.04 | 2.77 | 3.09 | 2.92 | 2.77 | 2.76 | .010 |
| 3.03 | 2.93 | 2.93 | 2.94 | 2.78 | 2.70 | .008 |
| 2.68 | 2.80 | 2.56 | 2.59 | 2.49 | 2.33 | .010 |
| 2.58 | 2.43 | 2.74 | 2.56 | 2.40 | 2.41 | .008 |
| 2.66 | 2.74 | 2.64 | 2.57 | 2.48 | 2.39 | .008 |
| 3.19 | 2.98 | 3.14 | 3.11 | 2.95 | 2.88 | .010 |
| 2.98 | 2.88 | 2.83 | 2.92 | 2.80 | 2.77 | .008 |
| 3.43 | 3.22 | 3.31 | 3.19 | 3.08 | 2.91 | .015 |
| 3.27 | 3.16 | 3.03 | 3.20 | 2.97 | 2.83 | .015 |
| 127 | 64 | 149 | 195 | 784 | 284 | |
| 294 | 129 | 287 | 336 | 1235 | 533 | |
| 416 | 247 | 426 | 426 | 1411 | 1318 | |

[c]1 = Very Good, 2 = Good, 3 = Fair, 4 = Poor, 5 = Very Poor.

$R^2$ values are for multiple regression analyses in which survey items were regressed on occupational groups (effects coded).

Respondents specifying occupation as "other" and respondents missing occupational codes are not included in the occupational categories in this table.

(promotions); and work is important and gives a feeling of accomplishment (intrinsic).

Table 2.4 presents the average ranking (calculated across all respondents) for each job characteristic in each year these questions were asked. As the table indicates, Americans in 1973 rated "intrinsic" aspects of work (i.e., having a job that gives them a feeling of accomplishment) as the job characteristic that they would prefer most in a job, followed in order by: promotions, income, job security, and hours. What is perhaps surprising is that this basic ordering, despite some minor fluctuations, has remained remarkably stable during the subsequent two decades: the rank order of these characteristics has remained virtually the same, with intrinsic aspects of work being the job characteristics most preferred by Americans in general and hours the least preferred. The relatively and consistently low emphasis that Americans place on job security (i.e., "danger of being fired") is somewhat surprising, given the fluctuations in unemployment

TABLE 2.4  Trends in Work Values in the United States, 1973-1994 General Social Survey

| Year | Average Ranking Across Categories of Values | | | | |
|------|--------|----------|-------|------------|-----------|
|      | Income | Security | Hours | Promotions | Intrinsic |
| 1973 | 2.72 | 3.60 | 3.99 | 2.61 | 2.08 |
| 1974 | 2.81 | 3.69 | 3.91 | 2.56 | 2.03 |
| 1976 | 2.67 | 3.53 | 4.10 | 2.62 | 2.08 |
| 1977 | 2.65 | 3.50 | 4.12 | 2.57 | 2.16 |
| 1980 | 2.59 | 3.73 | 4.14 | 2.55 | 1.99 |
| 1982 | 2.41 | 3.25 | 4.13 | 2.78 | 2.43 |
| 1984 | 2.64 | 3.54 | 4.27 | 2.54 | 2.00 |
| 1985 | 2.64 | 3.61 | 4.26 | 2.43 | 2.06 |
| 1987 | 2.49 | 3.61 | 4.17 | 2.58 | 2.16 |
| 1988 | 2.60 | 3.60 | 4.18 | 2.56 | 2.06 |
| 1989 | 2.57 | 3.68 | 4.26 | 2.56 | 1.93 |
| 1990 | 2.61 | 3.62 | 4.22 | 2.59 | 1.97 |
| 1991 | 2.56 | 3.61 | 4.19 | 2.65 | 1.98 |
| 1993 | 2.62 | 3.49 | 4.24 | 2.63 | 2.02 |
| 1994 | 2.55 | 3.43 | 4.26 | 2.67 | 2.09 |

rates (and thus of the probability of finding alternative employment) during this period.

Another source of data on values comes from the Work Trends™ survey (see Table 2.3), in which respondents generally reported that work provides a feeling of accomplishment (on average, 78 percent agreed or strongly agreed with this item), with only minor fluctuations between 1985 and 1996. Similarly, respondents answered positively to the intrinsic item, "My job makes good use of my skills and abilities" (on average, 71 percent agreed or strongly agreed with the item). For this item, however, there was a noticeable positive trend from 1985 to 1996 for most occupational groupings, particularly operatives, laborers, and technicians. In contrast, substantial negative trends between 1985 and 1996 were evident for the extrinsic work aspects of pay, benefits, and job security for all occupational groupings. On average, 54 percent of respondents rated their pay as good or very good in 1985 and 49 percent by 1996. Likewise, 72 percent rated their benefits as good or very good in 1985, dropping to 64 percent by 1996. Furthermore, respondents were substantially more positive about their jobs in 1985 (70 percent answered good or very good) than in 1996 (47 percent answered good or very good).

Although not directly comparable to the General Social Survey results due to differences in questions and response formats, there is nevertheless consistency in the findings from these two nationally representative surveys. Americans tend to evaluate intrinsic aspects of work more positively than extrinsic aspects. Furthermore, trends evident in the WorkTrends™ data on skills, pay, benefits, and job security help shed light on the General Social Survey results.

## Work Shifts and the Timing of Work

Work shifts and work time, perhaps more than any other dimension of work, are likely to be driven by demographic change. For example, the increased entry into the labor force of women with small children might be expected to increase the relative incidence of part-time work and of evening and night shifts, as parents juggle jobs and schedules. Popular accounts suggest that

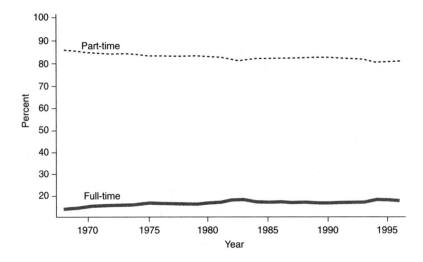

FIGURE 2.2 Percentage distribution of civilian employment by employment status.

there is a public perception of large growth in nonstandard work times and work shifts.

In fact, these changes are probably more modest than is commonly believed. Regarding the lengths of shifts people work, the percentage of the workforce that works part time did increase between 1968 and 1996, but much of the increase occurred in the 1970s (Figure 2.2).

Evidence on changes from 1973 to 1991 in the distribution of work time throughout the day reveals two main findings (Hamermesh, 1996). First, the proportion of hours worked at night (between 12 am and 4 am) has decreased steadily over this period. At the same time, the proportion of hours worked at the "fringes" of the work day (6-7 a.m. and 4-6 p.m.) increased over this period. These trends are similar for men and women, although more pronounced for men.

Another aspect of timing is the extent to which work is concentrated at particular hours of the day. To study this question, Hamermesh calculated the proportion of total work time worked during the eight most frequently worked hours. The data indicate small changes from 1973 to 1985 and declines for both men and women from 1985 to 1991. For men and women combined, for example, the proportion of work at these hours declined from .750 in 1985 to .730 in 1991, a difference that is statistically significant.

### Demographic Change Across Occupations

The aggregate occupation distributions of the workforce for selected years since the turn of the century, as well as occupation projections for 2005, are presented in Tables 2.5 and 2.6. Data for 1900 through 1980 come from decennial censuses; data for 1985-1995 come from the Current Population Survey and the Bureau of Labor Statistics. The major shifts over this century are well known and documented (see Table 2.5). In occupational locations, the

TABLE 2.5   Occupational Distribution of the Work Force, 1900-1995 (select years)

| Occupational Group | Year (percent) | | | | |
|---|---|---|---|---|---|
| | 1900[a] | 1930[a] | 1960[b] | 1970[c] | 1980[c] |
| Professional and technical | 4.3 | 6.8 | 11.4 | 14.6 | 16.0 |
| Managers and officials | 5.8 | 7.4 | 8.5 | 8.3 | 11.0 |
| Clerical | 3.0 | 8.9 | 14.9 | 17.8 | 19.0 |
| Sales | 4.5 | 6.3 | 7.4 | 7.2 | 7.0 |
| Craft and supervisors | 10.5 | 12.8 | 14.3 | 13.8 | 13.0 |
| Operatives | 12.8 | 15.8 | 19.9 | 17.8 | 13.0 |
| Laborers | 12.5 | 11.0 | 5.5 | 4.7 | 4.0 |
| Farmers and farm managers | 19.9 | 12.4 | 3.9 | 1.8 | 2.0 |
| Farm laborers and supervisors | 17.7 | 8.8 | 2.4 | 1.3 | 1.0 |
| Private household workers | 5.4 | 4.1 | 2.8 | 1.5 | 1.0 |
| Other service workers | 3.6 | 5.7 | 8.9 | 11.2 | 12.0 |

[a] Base population is "gainful workers."
[b] Refers to employed persons 14 years and older.
[c] Refers to employed persons 16 years and older.

SOURCES: U.S. Bureau of the Census. 1975. *Historical Statistics of the United States, Colonial Times to 1970, Bicentennial Edition.* Part 2. Washington, DC: U.S. Government Printing Office. Tausky, C. 1984. *Work and Society.* Itacca, IL: Peacock Publishers. Parts of this table were first published in Spenner (1988).

TABLE 2.6  Occupational Distribution of the Work Force, 1985-2005 (projected) (select years)

| Occupational Group | Year (percent) | | | |
| --- | --- | --- | --- | --- |
| | 1985[a] | 1990 | 1995 | 2005[b] |
| Executive, administrative, and managerial | 11.4 | 12.5 | 13.8 | 10.4 |
| Professional specialty | 12.7 | 13.7 | 14.5 | 15.5 |
| Technical and related support | 3.0 | 3.2 | 3.1 | 3.7 |
| Sales | 11.8 | 12.3 | 12.1 | 11.4 |
| Administrative support including clerical | 16.2 | 15.6 | 14.7 | 16.7 |
| Service | 13.5 | 13.5 | 13.6 | 17.2 |
| Precision production, craft, and repair | 12.4 | 11.5 | 10.8 | 10.3 |
| Operators, fabricators, and laborers | 15.7 | 15.1 | 14.5 | 12.4 |
| Farming, forestry, and fishing | 3.2 | 2.6 | 2.9 | 2.5 |

[a] All figures refer to employed persons 16 years and older.

[b] Projections refer to the "moderate" series estimates provided by the Bureau of Labor Statistics (i.e., under assumptions of moderate economic and industry employment projections).

SOURCES: Table A-22, Employed Civilians by Occupation, Sex and Age. 1991. *Employment and Earnings Statistical Abstract of the United States 1987 and 1995.* (U.S. Department of Labor, Bureau of Labor Statistics) 38(1):37. Silvestri, G.T. 1995. Occupational employment to 2005. *Monthly Labor Review* 118:60-87. Parts of this table were first published in Spenner (1988).

most dramatic change was the decline of farm-related occupations, from nearly 40 percent of the workforce in 1900 to less than 3 percent by 1995. The other major occupational shifts were major increases in the percentages of the workforce in professional, managerial, clerical, and service occupations from about 1 in 6 workers in 1900 to about 6 in 10 workers by 1980. Other major categories experienced more modest increases (sales and craft) or net decreases (laborers and private household workers). Operatives increased their percentage over the first half of the century but have gradually declined as a share of the workforce since 1960.

Table 2.6 provides the occupational distribution of the work force for more recent years, including the Bureau of Labor Statistics' projections for 2005 (Silvestri, 1995). All of the figures in Table 2.6 are based on the updated occupational classification system used by the Census Bureau. The projections shown here use the Census Bureau's "moderate" series estimates that assume moderate (versus higher or lower) assumptions about economic growth and industry employment projections. Managerial occupations, after peaking in the 1990s, are expected to decline in their relative share of the workforce, and the projection for professional specialty occupations shows steady growth in relative share over the period, which is expected to continue into the next century. Again, the demographic changes mentioned earlier may be among the underlying forces driving this change. Technical, sales, and administrative support occupations, although expected to grow in their absolute numbers, have and will retain about equal proportions of the work force over the period from 1985 to 2005. Service occupations, although fairly constant in relative share in the 1980s and 1990s, are expected to employ about 1 in 6 workers by 2005, a trend that is likely in part to reflect underlying demographic forces. Finally, occupations at the lower end of the socioeconomic spectrum—production, craft, repair, operators, fabricators, laborers, and farm, forestry, and fishing occupations—will all shrink in their relative shares of the workforce. The latter downward shifts are more likely to reflect technological and productivity changes rather than demographic shifts.

The long-term industry shifts are equally dramatic and generally reflect the well-known shift from jobs in goods-producing to service-producing categories. Service-producing industries contained 3 in 10 jobs at the turn of the century but had grown to nearly 7 in 10 jobs by 1980, and nearly three-quarters of all jobs in 1995.

A final set of general demographic trends that are important for understanding the evolution of occupational classification systems involves the gender composition of occupational employment. Long-term comparisons of trends in the sex segregation of occupations are difficult because of changes in occupational classifications systems and problems in comparability of categories over time (England, 1981; Beller, 1984; Reskin, 1993; Wootton, 1997). Occupational sex segregation is usually defined and measured as the percentage of women (or men) who would have to change occupations in order to be distributed to occupational categories in the same percentages as men (or women). The actual levels of sex segregation depend on the level of disaggregation in the categories, with more-detailed categories producing higher values of segregation. Furthermore, studies of detailed job categories show it is quite possible to have nearly complete sex segregation at the level of jobs even though aggregate occupational categories show modest or small levels of segregation of men and women.

Although there were some variations, the levels of occupational sex segregation appear to have been fairly stable, perhaps with small declines, over the first 60 to 70 years of this century. For example, the index of segregation computed on detailed occupational categories for the entire workforce declined from a value of 69.0 in 1910 to 67.6 in 1970 (that is, 67.6 percent of women in 1970 would have had to change their detailed occupation in order for women to have an occupational distribution that matched that of men) (Reskin, 1993). According to Jacobs (1998), by 1990 this index had been reduced to 56.4 percent, and in 1997 it reached 53.9 percent. He further notes that declines since the early 1970s have been the greatest for professionals and managers. The dissimilarity index for college graduates declined 20 points, compared with 11.8 points for those with less than a high school diploma.

## Demographic Change Within Occupations

To address the question of demographic change within occupations, the committee used data from the Current Population Survey for the 1983 and 1991 periods to analyze the changing demographic and skill composition of the workforce. For example, a particular workforce change over this period—such as the increase in the proportion of the workforce that is black—can be decomposed into across-occupation and within-occupation components.

This decomposition works because the total change in the proportion of blacks can be written as the sum of two components, each of which in turn is a sum. The first component is the sum over all occupations of the change in the proportion of the workforce in each occupation, multiplied by the (1983 or 1991) proportion of blacks in that occupation; this is the across-occupation component, since it is the change in the proportion of blacks that would have occurred had the proportion of blacks within each occupation remained constant, with only the occupational structure changing. The second component is the sum over all occupations of the change in the proportion of blacks in the occupation, multiplied by the proportion of the workforce in that occupation (in 1991 or 1983);[3] this is the within-occupation component, since it is the change in the proportion of blacks that would have occurred had the occupational structure remained unchanged. We carry out this analysis at the level of three-digit occupations.[4]

Table 2.7 reports the results of this analysis for demographic changes by age, sex, race/ethnicity, and educational attainment. Although not strictly demographic categories, we also present these decompositions by occupational and employer tenure and reported incidence of formal training, which provide further insight into within-occupation changes in the workforce or workers' skills. The first column summarizes the demographic and other shifts over this period. The demographic shifts, of course, correspond to the figures, showing increased representation of women of prime working age, workers with higher levels of educational attainment, and minorities. The last 6 rows indicate a decreased share with low (1 year or less) occupational tenure, and an increased share with high (10 years or more) occupational tenure. With regard to employer tenure, the data reflect increased shares at both the high and the low ends. Finally, a greater proportion of workers reported formal training either to improve skills on the job or to obtain the job.

The last four columns report on the decomposition, using alternative base years. Looking first at the age and sex changes, we

---

[3]The base year for the first component is the opposite of the base year for the second component; we report the calculation done both ways, although the results are qualitatively similar.

[4] This analysis required using a cross-walk between occupational codes used in the two years, provided by the Bureau of Labor Statistics, to handle the relatively small number of changes in these codes.

TABLE 2.7   Changes in Demographic Characteristics of the Workforce, Across and Within Three-Digit Occupations, 1983-2001

| | Total Shift | Proportion in 1983 | Proportion in 1991 |
|---|---|---|---|
| Proportion employed | | | |
| Women, 16-24 | −0.021 | 0.094 | 0.073 |
| Women, 25-54 | 0.043 | 0.285 | 0.328 |
| Women, 55-64 | −0.004 | 0.046 | 0.042 |
| Women, 65+ | 0.001 | 0.012 | 0.013 |
| Men, 16-24 | −0.025 | 0.104 | 0.080 |
| Men, 25-54 | 0.019 | 0.375 | 0.393 |
| Men, 55-64 | −0.012 | 0.066 | 0.055 |
| Men, 65+ | −0.001 | 0.018 | 0.016 |
| Black | 0.009 | 0.093 | 0.102 |
| White | −0.017 | 0.881 | 0.864 |
| Hispanic | 0.023 | 0.052 | 0.075 |
| < 4 years of high school | −0.040 | 0.184 | 0.143 |
| 4 years of high school | −0.014 | 0.367 | 0.353 |
| 1-3 years of college | 0.025 | 0.233 | 0.258 |
| 4+ years of college | 0.030 | 0.216 | 0.246 |
| 1 or fewer years in occupation | −0.023 | 0.191 | 0.168 |
| 1 or fewer years with present employer | 0.033 | 0.230 | 0.263 |
| 10 or more years in occupation | 0.058 | 0.337 | 0.395 |
| 10 or more years with present employer | 0.029 | .270 | 0.299 |
| Received formal training to obtain job[c] | 0.024 | 0.093 | 0.118 |
| Received formal training to improve skills[c] | 0.049 | 0.110 | 0.159 |

SOURCE: Demographic variables are from the NBER extracts from outgoing rotation group files of the CPS; training and tenure variables are from CPS training supplements. Black and white refer to race. Hispanic refers to ethnic origin.

[a]Using as weights 1983 proportion of occupation and 1991 proportion of demographic group within occupation.

| Percent Demographic Shift Across Occupations[a] | Percent Demographic Shift Within Occupations[a] | Percent Demographic Shift Across Occupations[b] | Percent Demographic Shift Within Occupations[b] |
|---|---|---|---|
| 9.96 | 90.04 | 4.18 | 95.86 |
| 6.78 | 93.22 | 10.28 | 89.72 |
| 27.58 | 72.42 | 18.47 | 81.53 |
| −30.93 | 130.93 | −52.58 | 152.58 |
| 11.54 | 88.46 | 8.03 | 92.01 |
| 21.46 | 78.49 | 5.46 | 94.54 |
| 1.65 | 98.35 | 7.01 | 92.99 |
| 27.78 | 73.02 | 38.89 | 61.11 |
| −7.95 | 107.95 | −9.45 | 109.45 |
| −2.27 | 102.27 | −0.93 | 100.93 |
| −6.49 | 106.53 | −10.55 | 110.59 |
| 26.61 | 73.39 | 22.63 | 77.34 |
| 64.44 | 35.56 | 69.51 | 30.42 |
| 13.26 | 86.74 | 8.92 | 91.08 |
| 56.18 | 43.82 | 56.90 | 43.14 |
| 10.09 | 89.91 | 13.90 | 86.10 |
| −10.56 | 110.56 | −14.69 | 114.69 |
| −2.11 | 102.13 | −2.35 | 102.35 |
| −1.94 | 101.97 | −1.36 | 101.36 |
| 20.20 | 79.80 | 21.39 | 78.61 |
| 15.84 | 84.16 | 18.02 | 82.00 |

[b]Using as weights 1991 proportion of occupation and 1983 proportion of demographic group within occupation.

[c]Formal training refers to company training programs, including apprenticeships. It does not include informal on-the-job training, in-school training in high schools, postsecondary schools, colleges or universities, or training in the armed services.

see that, with the possible exception of men age 65 and older, the changes were almost exclusively within occupations.[5]  As an example, the proportion of the workforce that is female and between ages 25 and 54 rose by .043 over this period; using the first decomposition, 93 percent of this change was within occupations.

This finding is even stronger for changes by race and ethnicity. The across-occupation shifts would have reduced slightly the proportions of blacks and Hispanics, as indicated by the negative values for the across-occupation component of the decomposition.  This implies that the occupations in which blacks and Hispanics tended to be represented have shrunk, so the increase in the proportion of the workforce in these categories was all within occupations.

The changes by educational attainment are more balanced. The decline in the proportion with less than four years of high school is largely a within-occupation phenomenon, as is the increase in the proportion with some college.  However, the smaller change (decline) in the proportion with 12 years of schooling occurred partly across and partly within occupations, whereas about half of the increase in the proportion with at least four years of college was across occupations.[6]

The decomposition for employer tenure, in particular, indicates growing diversity of the workforce within occupations, as the entire increase in the proportion both with low tenure *and* with high tenure occurred within occupations.  Finally, the decomposition for training also indicates possible changes within occupations, as approximately 80 percent of the increase in both types of training arose within occupations, pointing predominantly to changing skill requirements of existing occupations, rather than growth of higher-skilled occupations.

To summarize, this evidence on demographic and workforce

---

[5] A negative value for the across-occupation shift is not at all anomalous.  It arises when occupations in which a particular group is represented shrink, so that a within-occupation increase in the representation of this group is necessary simply to hold the overall representation of the group constant.

[6] This latter calculation must be regarded cautiously, given changes in the coding of education in the Current Population Survey between 1983 and 1993.

changes across and within occupations indicates that most of these changes in recent years have occurred within occupations. This does not necessarily imply that occupational classifications were increasingly challenged. But it does imply that there was increasing diversity of the workforce and requirements of workers within narrow occupational groupings. If this diversity is also associated with diversity in the content of work and the skills of workers, which seems particularly likely given the findings for tenure and training, then it is more likely that occupational classifications may have to be reevaluated.

## Changes in Wage and Earnings Inequalities Within Occupations

The other salient change in recent decades is the sharp increase in the dispersion of wages or wage inequality. Most economic research on this topic focuses on either within-group or between-group inequality, with groups typically defined by education, age, and experience. Thus, for example, between-group inequality may refer to the wage (or earnings) differential between workers with a high school diploma and workers with a college degree, whereas within-group inequality may refer to the dispersion of wages for those with a college degree. Recent research has clearly established that, for education, age, and experience groups, both between-group and within-group wage inequality has increased substantially over the past two decades (see, for example, Blackburn et al., 1990; Bound and Johnson, 1992; and Katz and Murphy, 1992). For example, wage differentials between college and high school graduates have widened, as has the dispersion of wages for high school students.

The rise in both components of inequality, especially the between-group component, has been linked to increased relative demand for workers with higher educational attainment, stemming in part, perhaps, from the diffusion of computers in the workplace, but due no doubt to other factors as well, such as intensified competition in global and domestic product markets, the decline of unions, and declines in the ratio of the minimum wage to average wages in the economy. To some extent, then, rising inequality has been attributed to changing demands for skills performed on the job.

This suggests that indirect evidence can be obtained on whether skills or tasks performed in jobs are changing by looking at changes in wage or earnings inequality within occupations. Research suggests that within detailed occupations, there is relatively little variation in educational attainments. Thus, an increase in inequality within detailed occupations is unlikely to be related to changes in wage or earnings differentials associated with education. Rather, such an increase in inequality is probably due, at least in part, to changing demands for particular skills performed by workers within an occupation.

Such changes are likely to indicate one of three things. First, workers may be entering the occupation who perform new skills or tasks not previously integral to the occupation. Second, the nature of the skills or tasks required of workers in the occupation may be changing, with those who can ably perform the newly required skills and tasks earning labor market rewards, and vice versa. Third, there could simply be growing dispersion in the types of workers (differentiated by skills and tasks) in many occupations.

In all three cases, however, the evidence of rising inequality within occupations would suggest that the specific or "typical" work performed within the occupation has changed. In contrast, if the increase in inequality is largely an across-occupation phenomenon, it may be related to the same factors that have changed wage and earnings differentials by schooling, experience, etc., but it would not provide indirect evidence of changing and more variable work within occupations.

Table 2.8 reports evidence on changes in wage and earnings inequality overall, within occupations, and between occupations. These changes are also calculated over the 1983 to 1991 period, which are both relatively high unemployment years, using the Current Population Survey. The first thing the table reveals is that earnings inequality has not risen, but wage inequality has risen (i.e., an increase in the log variance of .024 indicates a 2.4 percent increase in the variance). For the purposes of this investigation, wages are of more interest than earnings, as they reflect the price of a unit of labor and do not reflect changes in hours or

TABLE 2.8    Changes in Wages and Earnings Inequality Across and Within Three-Digit Occupations, 1983-1991

|  | 1983 | 1991 |
|---|---|---|
| Theil index of wage inequality | 0.1464 | 0.1530 |
| Within-occupation inequality | 0.0939 | 0.0967 |
| Between-occupation inequality | 0.0525 | 0.0563 |
| Theil index of earnings inequality | 0.1947 | 0.2000 |
| Within-occupation inequality | 0.1122 | 0.1162 |
| Between-occupation inequality | 0.0825 | 0.0838 |
| Variance of log wages | 0.3025 | 0.3267 |
| Average within-occupation variance | 0.1870 | 0.2061 |
| Between-occupation variance | 0.1155 | 0.1206 |
| Variance of log earnings | 0.6183 | 0.6132 |
| Average within-occupation variance | 0.3638 | 0.3747 |
| Between-occupation variance | 0.2545 | 0.2386 |

SOURCE: NBER extracts from outgoing rotation group files of the CPS. Earnings and wages are in real 1983 dollars and real 1991 dollars per hour. Real weekly earnings are truncated at 999 dollars per week in both 1983 and 1991.

weeks of work.[7]    The decompositions for wage inequality indicate that about 42 percent (using the Theil decomposition) to 79 percent (using the variance decomposition) of the rise in wage inequality occurred within detailed occupations during the period. As explained above, the increase in within-occupation wage inequality may point to changes in the types of work performed by workers in the standard occupations, possibly accompanied by growing dispersion within occupations in the skills and tasks required of workers.

## Summary of Demographic Findings

There are difficulties inherent in assessing how well current occupational classification systems accommodate a changing

---

[7] It is also possible that the use of the same relatively low nominal top-code for earnings in 1983 and 1991 depresses earnings inequality in 1991.

workplace. As a result, our analysis is somewhat speculative and perhaps contributes as much by posing research questions as by answering them. What we can establish is that the long-term increasing diversity of the U.S. workforce has been mirrored in occupations in recent years. In terms of both demographic characteristics and pay, workers in the occupations that we currently use to classify the workforce are increasingly diverse, making it more likely that such occupations include men and women, whites and blacks, more-tenured and less-tenured workers, and high-wage and low-wage workers. This increased diversity within occupations constitutes indirect evidence that the correspondence of current occupational classifications with the jobs that workers do is breaking down, since it is plausible that the increasing diversity of workforce characteristics and wages is reflected in increasing diversity of work.

Nonetheless, the evidence is limited in two important ways. First, it is indirect. It is possible, for example, that a more diverse workforce now does the same jobs that more homogeneous workers performed in the past, in which case this increasing diversity need not pose any challenge to occupational classification. However, at least from the perspective of economics, we would regard this latter possibility as far more plausible if there were not growing variance of wages within occupations; if wages ultimately reflect productivity, growing variance of wages is an indicator of growing variance of productivity, which in turn seems likely to be linked to increased diversity of work.

Second, we can document increasing variability in the type of work done within occupations as we now describe them. We do not attempt to assess the implications (presumably, the costs) of any failure of occupational classification systems to adapt to this increasing variability. On one hand, we could speculate that private businesses in a competitive environment find other ways to organize work to best utilize their workforce. On the other hand, occupational classification is also important in institutions that do not compete in the market—such as public organizations involved in training and career planning and the military.

## CONCLUSIONS AND IMPLICATIONS

The characteristics of the workforce and other features of the external environment reviewed here are changing in ways that will continue to affect the context, content, and outcomes of work. Some of these changes are readily tractable, such as changing demographic patterns. Others, such as changes in markets and technologies, are less easily measured or observed. However, in doing their work, those who design work structures and occupational analysis systems need a solid understanding of what is known and what existing research suggests, but cannot at this point document conclusively, about the effects of these changes. The evidence presented suggests the following conclusions:

1. Increased competition in product and financial markets will continue to exert pressures to hold down compensation costs, increase uncertainty over job stability, and call for emphasis on quality, innovation, and flexibility in work processes and outcomes.

2. Changing technologies will continue to alter skills and eliminate and create jobs at a rapid rate. Although skill requirements for some jobs may be reduced, the net effects of changing technologies are more likely to raise skill requirements and change them in ways that give greater emphasis to cognitive, communications, and interactive skills—points that are documented in more detail in Chapter 4.

3. Demographic changes will increase the diversity of individuals and groups across and particularly within occupations and organizations. Much more research is needed to understand the full implications of increased diversity, although the evidence available to date suggests that it will alter many of the social processes (communications, conflict, cohesion, etc.) that affect work outcomes. Changing workforce demographics will also alter occupational structures by increasing demand for goods and services that in the past were more often provided by family members who were not part of the paid workforce. Thus, managing diversity and addressing the consequences of more household hours being devoted to paid work will be increasingly

important tasks of both leaders and participants in teams, groups, and organizations.

4. One well-documented labor market outcome of recent decades has been an overall increase in the inequality of wages and incomes. For our specific purposes here, the variation in wages observed within occupations is especially important, since it is another indicator of how an outcome of seemingly similar work varies more today than in the past. Whether this trend continues, remains constant, or reverses in the future is a critical question worthy of study and active consideration both for its implications for work design and occupational analysis and for the broader aspects of work and employment policy.

# 3

# Changes in the Organizational Contexts of Work

C hanges in organizational strategies, structures, and processes are both affected by the external forces reviewed in Chapter 2 and in turn have significant effects on the nature of employment relationships, jobs, occupations, and occupational structures. In this chapter, we focus on the most noticeable of these organizational changes and their effects on work structures and content. We conclude by considering the implications of these changes for occupational analysis and classifications. In doing so, we highlight the organizational decisions that need to be factored more directly into systems for analyzing how work is done today and how it might be shaped in the future.

## ORGANIZATIONAL RESTRUCTURING

In this section, we discuss two important developments in the organization of firms. The first is downsizing and its implications for job security and job stability. The second is the trend toward flatter organizational hierarchies and the implications for how work is divided between managers and nonsupervisory employees. With flatter organizations, lower-level employees are asked to take greater responsibility as equal members of teams (National Research Council, 1997a).

## Downsizing: Changes in Job Security and Job Stability

Downsizing, which refers to reducing the size of the workforce, was used by many firms in corporate America during the late 1980s and early 1990s to respond to new financial exigencies. More recently, however, survey results obtained by the American Management Association from its member companies show that workforce reductions are increasingly strategic or structural in nature rather than a response to short-term economic conditions associated with declines in business (American Management Association, 1996:2). The 100 largest companies in the United States reported that 22 percent of their workforce had been laid off since 1978, and 77 percent of those cuts involved white-collar jobs. Approximately 23 percent of companies surveyed in 1997 reported outsourcing as a cause for downsizing.

In recent years, firms are hiring at the same time that they are laying off: 31 percent of the large, traditional employers surveyed by the American Management Association in 1996 were adding and cutting workers at the same time, and the average firm that had a downsizing in fact was growing by 6 percent. Smaller firms were creating more jobs than they were cutting, whereas large firms with over 10,000 workers were more likely to see actual declines in overall employment. The telecommunications industry provides a good illustration. At its peak, AT&T employed 950,000 people and around 450,000 after divestiture. Now it employs around 250,000 with plans to cut further. Yet the telecommunications industry still has about the same number of employees as a decade ago. As AT&T has shrunk, through both divestitures and downsizing, smaller competitors have grown up around it (Nocerra, 1996).

Looking at data from the labor market on workers who are displaced from their jobs, Farber (1997) finds that the overall rate at which workers have been permanently displaced backed down a bit in the 1980s from the peak of the recession period, 1981-1983, but then rose again—despite the economic recovery—and jumped sharply through 1995. The rate at which workers lost their jobs was actually greater in 1993-1995, a period of significant economic expansion and prosperity in the economy as a whole. It is difficult to think of more compelling evidence that the nature of the employment relationship has changed than this.

Perhaps the most telling change in the employment relationship has been the reduction in white-collar and management jobs, which were traditionally the most protected. The 1996 American Management Association survey finds that, although salaried employees held roughly 40 percent of all jobs, they account for over 60 percent of all the employees cut. The number of supervisory employees eliminated as a percentage of all employees that were cut doubled between 1990-1991 and 1993-1994 to 26 percent (Rousseau and Anton, 1991). The rate of displacement was actually higher for managers in the 1980s than for other occupations, controlling for other characteristics (Rousseau, 1995). It rose sharply through the early 1990s but appears to have declined somewhat from 1993-1995 (Farber, 1997), perhaps simply because of regression to the mean.

The "churning" of the workforce—hiring and laying off at the same time—had the biggest negative effects on middle management: three jobs were cut for every one created. These are the positions that are the most entrenched within the internal employment system. Professional and technical jobs, in contrast, benefited from it: five new jobs were created for every three cut. These jobs have the skills and responsibilities that translate easily across organizations. The changes associated with churning in some cases go beyond simply rearranging which employees hold permanent jobs. A survey of 500 human resource executives whose companies had downsized found that a third refilled at least some of the positions that had been cut but that 71 percent did so with either temporary or contract workers (Lee Hecht Harrison, 1997).

But large employers of the kind represented in the American Management Association surveys account for only a modest percentage of total employment in the economy. Many workers were never employed in internal labor markets with prospects of job security. During the course of the 1980s and especially the 1990s, a widespread perception developed that the employment relationship had changed in fundamental ways. A steady drumbeat of stories in the media fed the perception that employers were less attached to their employees. The overall theme is summed up by an article in *Time* (Church and Greenwald, 1993) suggesting that Americans are realizing that the great American job is gone and that they should forget any idea of career-long employment with a big company. These perceptions affected employees' expectations about what it means to work in an organization and the role of loyalty to the employer.

Empirical research on changes in the employment relationship for the workforce as a whole suggests a more temperate view. There are some changes evident in the data, but some empirical measures of the employment relationship do not point to changes. When changes do appear, they are not overwhelmingly large. Furthermore, evidence of a long-term trend toward declining attachments in the workforce as a whole is modest and appears to be concentrated in the early 1990s, the most recent past. This raises the question of whether we are witnessing a period of short-term restructuring that may portend little about the future. It is possible, for example, that the evidence so far of modest change is driven by rather significant changes by large corporations, balanced by stability in the relationship for the majority of the workforce employed under other arrangements.

Empirical studies of the employment relationship based on large, random samples of the population and workforce available over a long span of years have focused primarily on two salient features of the employment relationship: job stability and job security.

## Job Stability

Job stability is defined in the recent literature as a measure of how long jobs last (or how quickly they end), irrespective of the source of turnover. Changes in job stability are driven either by voluntary quits or by employer-initiated separations such as layoffs or firings. Employer-initiated separations tend to be greater and employee quits fewer during recessions (because most employees who quit do so to take jobs elsewhere that are scarce in recessions), offsetting tendencies reflected in the overall measures of job stability.

Diebold, Neumark, and Polsky (1997), examining the temporal evolution of job stability in U.S. labor markets over the 1970s and 1980s, used data assembled from a sequence of Current Population Survey (CPS) tenure supplements. They use these CPS supplements to construct artificial cohorts, with which job retention rates (i.e., the probability of retaining one's job) over various spans of years can be computed and compared over time. This is preferable to looking at distributions of incomplete tenure spells at a point in time, which can be affected by, among other things, new entrants to the labor market, and hence obscure information on how long jobs are lasting.

Through the 1980s to January 1991, they found little or no change in average job stability for the workforce as a whole in the U.S. economy (see Table 3.1). Columns (1), (2), and (4) in Table 3.1 summarize these results for 4-year retention rates. Columns (1) and (2) report the estimated rates, and column (4) reports the changes. As shown in the first row, the overall change is a very slight decline of .002. As shown in the other rows, at a disaggregated level, job stability fell for those with 2-9 years of tenure and rose for the least and especially the most tenured workers (15 or more years of tenure). By age group, job stability fell only for the youngest workers. Column (7) reports changes in estimated 10-year retention rates, over a longer sample period. Here, too, the overall change is negligible (–.001), although the pattern by age and tenure group is somewhat reversed.

Columns (3), (5), and (6), as well as column (8), report newer results from Neumark et al. (1997), updating the estimates through 1995 (Wyatt Company, 1995). The estimates reported in the table reflect a relatively consistent picture: first, 4-year retention rates still show essentially no overall change, if anything rising slightly compared with earlier 4-year spans. However, in the 1991-1995 period, job stability declined rather substantially for workers with 9 or more years of tenure, a finding that may reflect the types of stories that have appeared in the media regarding changes in job stability for "career" employees. In the 8-year retention rate results reported in column (8), similar (although less sharp) declines for more-tenured workers appear. Moreover, for the period considered, 8-year retention rates fell modestly overall, by .021. Finally, the results disaggregated by age indicate nontrivial declines for all age groups except those age 55 and over. This suggests that conditional on age, job stability has declined

TABLE 3.1 Changes in Job Retention Rates

| | Changes in 4-year retention rates 4-year retention rates | | | Rate comparisons | | | Changes in 10-year retention rates | Changes in 8-year retention rates |
|---|---|---|---|---|---|---|---|---|
| | 83-87 | 87-91 | 91-95 | 83-87 87-91 | 87-91 91-95 | 83-87 91-95 | 73-83 to 81-91 | 83-91 to 87-95 |
| | (1) | (2) | (3) | (4) | (5) | (6) | (7) | (8) |
| All workers | .539 | .536 | .544 | -.002 | .008* | .005 | -.001 | -.021* |
| **Tenure groups** | | | | | | | | |
| 0 to < 2 years | .329 | .346 | .391 | .016* | .045* | .061* | -.004 | .031* |
| 2 to < 9 years | .586 | .548 | .564 | -.038* | .016* | -.021* | .012* | -.052* |
| 9 to < 15 years | .827 | .816 | .748 | -.010 | -.069* | -.079* | .063* | -.047* |
| 15 or more years | .630 | .702 | .633 | .072* | -.070* | .003 | -.043* | -.020* |
| **Age groups** | | | | | | | | |
| 16-24 | .305 | .281 | .292 | -.025* | .011 | -.014 | .040* | -.030* |
| 25-39 | .585 | .577 | .573 | -.009 | -.004 | -.012* | -.005 | -.037* |
| 40-54 | .686 | .683 | .674 | -.004 | -.009 | -.012 | -.044* | -.029* |
| 55 and over | .469 | .468 | .451 | -.000 | -.018 | -.018 | -.005 | .005 |

Estimates are adjusted for heaping of tenure responses at multiples of 5 and 10 years, rounding, changes in tenure questions, and the business cycle. Estimated changes statistically significant at the 5-percent level are indicated with an asterisk.

SOURCE: Diebold et al. (1996, 1997) and Neumark et al. (1997).

more, while the overall decline has been moderated by the effects of an aging workforce coupled with higher job stability experienced by older workers (through age 55). Thus, the evidence of declines in job stability is a bit stronger if we compare the experiences of similar age groups over time than if we consider the workforce as a whole. Furthermore, when examining these changes by occupational group, the results suggest that managers have experienced a significant decline in job stability in the 1990s (Neumark et al., 1997).

Some studies have reported sharper evidence of declines in job stability (e.g., Swinnerton and Wial, 1995; Rose, 1995; Boisjoly et al., 1994; Marcotte, 1996). However, these findings are largely artifacts of changes in survey methods or errors in classifying workers (see, e.g., Diebold et al., 1997; Polsky, 1996; Schmidt and Svorny, forthcoming). In addition, the last two of these studies that report declines measure year-to-year separations, so they could be detecting more turbulence for those on new jobs, and the probability of a long attachment between workers and firms may not have fallen.

Overall, then, there is some evidence that average job stability declined modestly in the 1990s. The relatively small aggregate changes mask sharper declines in stability for workers with more than a few years of tenure. These sharp declines are partially offset in the aggregate by gains in job stability for low-tenure workers at the beginning stages of attachment to an employer. The changes by tenure group contrast with the 1980s, and along with the declining stability of managers are more consonant with the increase in job loss among "career workers" described in the popular press. However, because these changes appear principally in recent years, it is difficult at this point to know whether there has been a permanent change in job stability.

## Job Security

In contrast to the preceding section, the concept of job security is presented from the perspective of the employee and focuses on involuntary terminations. Since voluntary terminations are likely to be beneficial to workers, whereas involuntary terminations are likely to be costly to them, a shift in the composition of terminations toward involuntary ones could help explain the perception of increased anxiety over job loss.

A number of studies show evidence of increased involuntary job loss (*Fortune*, 1994; Valletta, 1996; Polsky, 1996; Boisjoly et al., 1994; Farber, 1996). This evidence appears to be consistent with changes in the employment relationship toward a less secure relationship. The data reported in Table 2.3 of the previous chapter echo this trend. The job insecurity of U.S. workers increased between 1985 and 1996 for all occupational categories, with most of the increase occurring in the 1990-1996 period for most categories. However, some caution is in order regarding both the interpretation of the evidence and the evidence itself. The fact that the composition of job separations has shifted from voluntary quits to involuntary separations does not necessarily imply the disappearance of long-term jobs. Just as reasonably constant job stability does not necessarily mean no change in job security, declines in job security do not necessarily imply reductions in job stability. It simply means that the employer—as opposed to the employee—is now more likely to be the one ending the relationship.

The fact that employees who want long-term relationships may no longer have the same possibilities as in the past is a change of some importance, however. Although workers are being involuntarily terminated at a higher rate, they are also voluntarily staying on the job longer, perhaps because there are fewer alternatives elsewhere. This trend is supported by the data shown in Table 2.3. They reveal a large decrease, between 1985 and 1990, in intention to leave current employment for all occupational categories and little or no change between 1990 and 1996. Valletta (1996) suggests that reduced quits may reflect increased insecurity, as adverse labor market developments make workers unwilling to cut their ties to their existing employers and try their luck on the market. It is also possible, however, that many workers are finding their present employers relatively more attractive than in the past—whether because of higher wages, better long-term prospects, or other factors—and as a consequence are quitting less. Finally, changes in the state of the labor market, particularly the overall rate of unemployment, may moderate some of the trend toward greater involuntary separations.

Although there are some ambiguities in the findings, on balance the evidence points to recent declines in job stability for at least some groups of workers during the first half of the 1990s. How fundamental these changes are depends on whether they reflect the beginning of a long-term trend, a question that can best be addressed by examining other aspects of the employment context.

## Flatter Hierarchies:  Changes in Management and Teamwork

Changes in the organization of work have made it possible to eliminate many jobs from organizations and may also facilitate their rapid restructuring. We examine two such changes: changes in managerial jobs and the rise in team-based work.

### Managerial Jobs

Managerial jobs may well be the ones that have experienced the greatest transformation in their structure in recent years, and their transformation has been the subject of considerable attention from the business press. To understand what has changed, it is sensible to start by describing the characteristics of the "old" internal labor market of managers.

The distinctive characteristic of traditional managerial careers is the existence of a promotion ladder and the notion of moving up it (Rosenbaum, 1984; Baker et al., 1994). Ladders of upward mobility were well defined. One implication of these ladders is that most movement was in the upward direction. Managers either were promoted or stayed where they were. Lateral shifts were very rare; demotions were essentially unheard of.

All observers agree that managers did not face serious risks of layoffs until the late 1970s (MacDuffie, 1996). Nor was pay at risk. Senior managers began the 1980s with the vast majority of their pay package in the form of base or guaranteed pay that did not vary with either their performance or that of the organization (Useem, 1996).

Another aspect of managerial employment systems was the strong bias toward continually increasing the proportion of managerial jobs in the workplace. In 1929, American manufacturing firms employed 18 administrative workers for every 100 employees; by 1950 the figure was 24; by 1960 it was 29; and in 1970 it was 30 (Gillen, 1994:66, 82). In part this reflected the impact of increased organizational size and the planning requirements. But American firms as a group are also more managerially intensive than are companies in other developed nations (Gorden, 1994). There are a number of explanations for this pattern. The most consistent finding in the executive compensation literature has been that managerial salaries in compensation systems based on the U.S. model rise with the size of the unit reporting to the manager (see, for example, Milgrom and Roberts, 1992).

All of these characteristics of managerial employment seem to have changed dramatically in recent years. Press accounts of downsizing in large corporations suggest that managerial jobs are at substantial employment risk. This impression is supported by evidence from the Dislocated Worker Surveys, which suggests that the risk faced by managers was actually greater than for other employees, other things being equal (see below). Case study evidence also supports this conclusion.

Managerial compensation has also changed radically in the direction of variable, at-risk pay. The fraction of managerial pay that is contingent on performance has increased (Useem, 1996). The percentage of managerial compensation associated with bonus payments rose from 1986 to 1992, as did the returns to skill (measured by job evaluation scores: Scott et al., 1996). Risk has gone up as firms seek to use compensation to align more tightly the behavior of managers to the presumed goals of shareholders.

For most managers, upward mobility is much less probable than it used to be. The use of task or project teams that come together for specific, temporary tasks and then disappear replaces permanent bureaucratic hierarchies and, in the process, reduces the number of levels in the organizations. The use of teams among production and first-line workers sharply reduces the need for supervisors and the managers whose jobs had been to manage the supervisors. These changes have removed several rungs from promotion ladders, and case studies of a variety of firms suggests that the chances of managers to move up have diminished (Batt, 1996; Scott et al., 1996). In some industries, such as the high technology community of Silicon Valley, networks of companies seem to have replaced the large, vertically integrated organization. Managerial and professional employees in these communities advance by moving across organizations rather than moving up within them (Saxenion, 1994).

## Teams, Teamwork, and Team-Work

The rise of team-based work structures is perhaps the most ubiquitous change affecting the workplace in the past two decades. The shift from individualized work structures to teams spread from production work to the executive suite and every area in between, forcing revisions in compensation systems, organizational charts, and corporate culture. The increased use of teams and teamwork appears to have implications for a number of issues related to jobs, occupations, and occupational structures. The most obvious of these are that teamwork broadens jobs and blurs boundaries between them. Teamwork also flattens hierarchies within organizations, as authority moves from what had been managerial and supervisory positions to the team. As a result, job ladders must be redefined as promotion prospects inside organizations erode.

The benefits of using teams in firms, laid out as far back as the 1920s by the well-known management observer and author Mary Parker Follett, include decisions that are better informed, greater commitment by employees to the work being performed, and a reduced need for formal supervision and administration (Graham, 1995). However, in the next several decades (1930s and 1940s), little real-world emphasis was given to the deliberate use of teams in organizations, although academic scholars were doing considerable basic research on topics related to group dynamics. Since the 1980s, some attention has been given in industry to the use of specialized teams (such as Scanlon Plan Teams and Quality Circle Teams), and a small number of innovative companies were starting to make greater use of teams in their organizations. In the 1990s, the popular business press has been replete with articles about the increasing use of teams in corporate America, often with a tendency to exaggerate their positive effects.

When a traditional workplace with narrowly defined jobs is replaced with a cross-functional team, the low-skilled, entry-level positions are often eliminated, and their tasks (such as housekeeping) added to those of existing jobs. The supervisory positions are also reduced, and their tasks are dispersed to the group as a whole. When employees are cross-trained to handle many different tasks, their skill sets shift to include a much broader set of knowledge, skills, and abilities. The hierarchy of jobs based on increasing skill within an area gives way to greater commonalities in status and more informal authority structures.

In the context of such teams, exactly how new employees learn the skills for this cross-functional team is less obvious than in the old workplace, where unskilled workers could be assigned to unskilled jobs and then slowly learn new skills and positions. There may be less mobility in the new team model, particularly when jobs are organized according to principles of scientific management, which has broadened individual jobs by eliminating narrow, hierarchical occupational structures. Employees in these new cross-functional teams may be much less likely to identify with a skill and the job defined around it, such as crafts (e.g., "machinist") and more likely to see their identity as a member of a specific group of employees. Variations in the nature of teams can lead to very different work structures and, in turn, different job and occupation descriptions. These variations further increase the heterogeneity in the workplace.

Much of what we know about teams comes from experience with production workers in manufacturing, on which the majority of the research has been done (Osterman, 1996; Hunter, 1998a; Batt, 1999a). Many of the benefits of teamwork appear to be applicable elsewhere (e.g., Cohen and Bailey, 1997; Batt, 1999a), and the use of teams may well continue to expand. The reasons for this include: the increasing complexity of certain types of work, which makes it difficult for individuals to perform it on their own, the tendency for organizations to attempt to flatten their structures by reducing their levels of management, and the related apparent trend to provide employees greater opportunity for decision making at lower organization levels. Support for the latter observation can be found in the WorkTrends™ data summarized in Table 2.3. Respondents in all occupational categories except sales reported somewhat greater participation in decision making in 1996 than in 1990 ("In my company, employees are encouraged to participate in making decisions which affect their work.").

There is fairly convincing evidence that teams can be used effectively in a wide variety of organizations and work contexts. Research using the National Establishment Survey finds that teams are more prevalent in manufacturing, even controlling for size of establishment, although self-managed teams that take over much of the day-to-day supervision of the work appear more common in services (Osterman, 1996). Other research also finds them used most often in manufacturing firms, somewhat less in service firms, and least in the government/public sector. Industry-specific studies find that the use of teams is still uncommon in industries such as retail banking or services (e.g., Hunter, 1998a; Batt, 1998b).

The business press suggests that teams became common during the 1990s and that their influence was increasing. For example, a 1993 article in *Personnel Journal* (Caudron, 1993) reported the results of a survey conducted by Development Dimensions International that showed that 27 percent of corporate respondents reported that their organizations use self-directed teams, and that 50 percent of the respondents predicted that the majority of their workforce would be organized in teams within the next five years. Likewise, a report prepared by the Towers Perrin consulting firm (in conjunction with IBM) in 1993 (cited in Tannenbaum et al., 1996) indicated that, in a study of 3,000 managers and executives from 12 different countries, respondents viewed the use of teams as an important competitive advantage in coming years and rated teamwork as one of the highest business priorities for the year 2000.

The best real evidence of the incidence of teams in the workplace comes from national surveys conducted in conjunction with the General Accounting Office (GAO), by the Bureau of the Census (the National Employer Survey [NES] from the National Center on the Educational Quality of the Workplace), the Bureau of Labor Statistics, and by independent researchers (the National Establishment Survey). These surveys tend to corroborate the business press conclusions, suggesting rapid increase in the use of teams in the workplace into the early 1990s (Cappelli et al., 1997; Gittleman et al., 1998).

Additional issues associated with teamwork also appear to have important implications for occupations and classification systems but have not yet been examined with solid empirical research. They include the following:

• How the nature of supervision changes, especially when teams are self-managed;
• The changing role of the middle manager;
• The relative power of functional and project managers;
• Implications for organizational structure and design;
• Implications for assessing team performance and reward/compensation structures as organizations shift from work structures based on individuals to teams; and
• Implications for career development, especially associated with shifting from narrow, functional work structures to more cross-functional teams.

## CHANGING EMPLOYMENT RELATIONSHIPS

Organizational restructuring sets off a chain reaction of additional changes in the terms and conditions of employment. In this section we review the most important of these changes.

### Changes in Career Patterns

#### Careers Across Employers

Changes in business may be weakening the prospects for lifetime careers within a single company, but other developments are increasing the opportunities for different career patterns that span organizations—what Arthur and Rousseau (1996) have called "the boundaryless career." These developments include the rising tide of mergers and acquisitions (Cappelli, 1999), alliances across companies (Lewis, 1995), and joint ventures (Harrison, 1994; Lorange and Roos, 1992). Such activities can give employees access to very different labor markets and networks of information and colleagues. There is evidence from employers that their recruitment and selection priorities have changed and that the trend, even among large companies, is toward hiring employees with experience for what have been seen as entry-level management jobs and away from hiring new entrants straight from college (Rynes et al., 1997).

The operation of firms through informal supplier networks, formal joint ventures, and similar arrangements has become the norm in some high-tech industries, especially those associated with Silicon Valley and similar industrial communities like Austin, Texas. Descriptions of the operations of these companies suggest that the flow of employees across companies is a distinctive characteristic of their operations, providing the crucial source for the transfer of technology and ideas, learning by employees, and flexibility in terms of a company's ability to restructure and change. For employees, this model of how a firm is organized makes it almost necessary to move across employers in order to progress in a career, given that the relatively flat structure of firms provides little prospect for internal promotion and that training and internal development have been replaced by hiring from outside as a means for meeting skill needs (Saxenion, 1994). The growth of outsourcing and greater reliance on the supply chain outside the firm (Lewis, 1995) in other industries (Gereffi, 1994) creates similar opportunities for employees to get exposure to information and opportunities outside their firm.

### Attachment to Occupations

Especially for managerial, professional, and technical employees, career paths across firms may become more common as employees move across organizations to perform similar work. If so, it is important to consider their attachment to occupations. Rose (1995), presenting limited information on the stability of employment within occupations in the 1970s and 1980s, found some evidence of modest increases in the length of time that individuals are remaining in the same occupation. Ultimately, this question calls for an analysis as detailed as that applied to job stability and job security.

Although an analysis based on retention rates is required to draw firm conclusions, Table 3.2 nonetheless reports some preliminary calculations based simply on the distributions of incomplete spells of occupational tenure in the 1983 and 1991 Current

TABLE 3.2 Changes in Employer and Occupational Tenure

| | Employer Tenure | | Occupational Tenure | |
|---|---|---|---|---|
| | 1983 | 1991 | 1983 | 1991 |
| | (1) | (2) | (3) | (4) |
| Mean | 7.05 | 7.12 | 8.54 | 9.07 |
| | (.04) | (.04) | (.04) | (.04) |
| Median | 4 | 4 | 5 | 6 |
| Proportion in range | | | | |
| 0-5 | .54 | .53 | .45 | .41 |
| 5-10 | .19 | .18 | .21 | .20 |
| 10-15 | .11 | .12 | .13 | .15 |
| 15-20 | .06 | .07 | .07 | .09 |
| 20+ | .10 | .10 | .13 | .14 |
| Mean by age | | | | |
| 16-24 | 1.70 | 1.54 | 2.09 | 1.94 |
| | (.02) | (.02) | (.02) | (.02) |
| 25-34 | 3.97 | 4.12 | 5.10 | 5.51 |
| | (.03) | (.03) | (.03) | (.04) |
| 35-44 | 7.29 | 7.55 | 9.00 | 9.74 |
| | (.06) | (.06) | (.06) | (.06) |
| 45-54 | 11.60 | 11.27 | 13.57 | 13.90 |
| | (.10) | (.10) | (.11) | (.10) |
| 55+ | 15.46 | 14.45 | 18.39 | 18.24 |
| | (.14) | (.16) | (.15) | (.17) |
| Median by age | | | | |
| 16-24 | 1 | 1 | 2 | 1 |
| 25-34 | 3 | 3 | 4 | 5 |
| 35-44 | 5 | 5 | 8 | 9 |
| 45-54 | 10 | 10 | 12 | 13 |
| 55+ | 13 | 12 | 16 | 17 |

Estimates are computed from 1983 and 1991 CPS supplements. All estimates are weighted. Standard errors of means are reported in parentheses.

Population Surveys (these estimates are based on data from Neumark et al., in press). For purposes of comparison, similar estimates for tenure with the employer are also reported. The figures in columns (3) and (4) indicate that over this period occupational tenure rose, with the mean rising by about one-half year and the median rising by about one year. This has occurred via a rather sharp decrease in the proportion with 0-5 years of occupational tenure, and increases in the proportion with 10 or more years of occupational tenure. The bottom rows of the table indicate that mean tenure has risen for all age groups except the oldest workers, and the median tenure has risen for all but the youngest group. Overall, then, although this evidence is very preliminary, it suggests that occupational tenure has not declined, and that workers are, if anything, remaining in their occupations somewhat longer.

### Networking Opportunities

If managerial and skilled employees in particular find better prospects in careers that span organizations, they must somehow develop mechanisms for getting information about opportunities elsewhere and for securing new positions. Sociological research has shown that networks are a major way that people get jobs. These networks are organized in terms of family members for some, but are also organized along dimensions such as religion, ethnicity or race, and so on. Granovetter, for example, found that 56 percent of the professional, technical, and managerial workers in his study used contacts to find their jobs. Of these, about a third used family-social contacts (defined as a relative, friend of the family, or social friend) to do so. Granovetter, using data from the Panel Study of Income Dynamics (Hill, 1992), concludes that almost half of the adults, regardless of ethnicity or gender, found their current jobs via friends and relatives. This proportion is confirmed by data from Lincoln and Kalleberg's study (1990) of the employees of 52 manufacturing plants in Indianapolis.

Family networks are especially important for groups that might otherwise be excluded from labor markets for good jobs, such as immigrants. Roger Waldinger's studies (cited in Granovetter, 1995) illustrate the importance of immigrants' networks in helping their members get jobs in New York City. Ethnic and racial groups also differ in their use of family ties: for example, analyses from the Multi-City Study of Urban Inequality showed that Hispanics benefit more from family ties in getting jobs than blacks or whites.

Trade unions and professional associations might also provide alternative mechanisms for employees to use both as sources of information and job mobility and as mechanisms for defining their identity in the workplace. Craft unions have long served those functions and have helped to structure labor markets in many industries, such as construction, through the creation of hiring halls and other mechanisms. The much greater involvement of unions as champions of training programs for their members, especially retraining programs, is a more contemporary innovation. The rise of "nickel funds" and other arrangements for funding joint union-management training programs has given employees more influence over the content of training programs and more choice in selecting them. Especially in companies in which job opportunities for union members are in decline, these programs provide mechanisms for changing careers through the acquisition of new skill sets that are often unrelated to current jobs (Ferman et al., 1990).

The decline of union coverage in the private sector has reduced the ability of unions to serve as an alternative source for training, a network for jobs and career mobility, and more generally for information and support when changing jobs. Nowhere is this decline more noticeable than in such industries as construction, in which unions not only helped organize the labor market but also developed elaborate mechanisms for acquiring and certifying the skills of workers through apprenticeship systems.

Professional associations may perform similar functions for managerial and white-collar jobs. Although the evidence is only anecdotal, it does appear that these associations are spending more of their energies on training and career advancement (including the traditional practices of networking) in response to the greater interest of their members in careers that span employers. The training and especially the certification aspects of these new efforts seem oriented toward producing something like a professional occupational model for jobs that had previously been seen as internal management positions, such as the new title of "certified employee benefits specialist."

Other arrangements, which include community-based placement services, partnerships between private staffing companies and state employment services, and other local networks, help workers to make matches with available jobs and to smooth the transition period between jobs (Carre and Joshi, 1997). For the most part, however, these institutions remain local and cover only a small percentage of the economy even in the communities where they operate. National efforts to develop labor market institutions, such as the National Skills Standards Board, which was originally envisioned as mapping out common job requirements and standards for credentials for virtually all jobs in the economy, have stalled.

## Nonstandard Work

The standard work arrangement that dominated the attention of policy makers, practitioners, and researchers in the United States since World War II was one in which persons worked full-time for a single employer who controlled their work, advanced steadily up a job ladder, and received a pension upon retirement. Evidence suggests that this traditional or standard arrangement is becoming less common, though perhaps it was never as prevalent as its impact would otherwise suggest. Replacing it are a series of nonstandard work arrangements such as regular part-time employment, working for a temporary help agency, working for a contract company, on-call work or day labor, independent contracting, and other forms of self-employment.

In 1995, 29.4 percent of U.S. labor force members worked in nonstandard work arrangements as defined by these categories, with 34.4 percent of female workers and 25.4 percent of males working in nonstandard jobs (see Kalleberg et al., 1997). Regular part-time workers constituted the majority of persons in nonstandard work arrangements (21.3 percent of women and 7.1 percent of men working in nonstandard jobs) and has been relatively constant for some time, followed by self-employed persons, including independent contractors (8.5 percent of women and 13.4 percent of men working in nonstandard work arrangements). The extent to which U.S. organizations use various flexible, nonstandard staffing arrangements is greater than the proportion of persons who work in such jobs: a recent survey (Houseman, 1997) found that 46 percent of U.S. organizations use workers obtained from temporary help agencies, 43.3 percent use contract workers, 38 percent use short-term hires, 27.3 percent use on-call temporary workers, and 71.6 percent use part-time employees.

Evidence on *changes* in employer practices over time is difficult to come by, although anecdotal evidence seems to suggest an expansion of contract workers and part-time workers in particular. Data on employees over time, however, suggests that the change in nonstandard work may be less dramatic. Part-time employment has not increased as fast since 1970 as earlier; the increase from 16.4 percent of the labor force in 1970 to 18 percent in 1990 is due almost entirely to the growth of involuntary part-time employment (Tilly, 1996). Self-employment has also remained relatively constant at 12 percent of the labor force over the past two decades.

Evidence for growth in independent contracting and other forms of contract work is suggested by the increase in the number of persons reporting income only as self-employed or independent contractors (Callaghan and Hartmann, 1991) and the growth in business and personal services (U.S. Department of Labor, 1995). Although contract work may represent an important change for an organization, it may not represent a decline of standard work for employees. (Consider, for example, a corporation that contracts out tasks previously performed by permanent employees. If the work goes to an organization with its own permanent employees, it may represent no net change in the proportion of standard jobs in the economy, even though it does represent a decline in job security of its current workers.) The most important increase in nonstandard work has been the rise of temporary help. As defined by employees of temporary help agencies, it has been growing at an average rate of 11 percent per year since 1972. But it still only accounts for about 3 percent of the labor force (Segal and Sullivan, 1996). Other means of using temporary help, such as an employer creating its own pool of temps, might put the figure substantially higher.

Nonstandard work arrangements are often associated with

jobs that are "contingent," i.e., are insecure in the sense of being of limited duration. Analyses of data from the February 1995 Current Population Survey (see Kalleberg et al., 1997) suggest that about 18 percent of men and women in nonstandard jobs (compared with only 5.4 percent of men and women in regular, full-time jobs) have jobs that are contingent. Nonstandard workers also tend to receive lower wages than regular full-time workers with similar personal characteristics and educational qualifications, although the extent of these wage gaps shrink once occupation, industry, and union status are taken into account. Workers in nonstandard work arrangements are also much less likely to receive fringe benefits, such as health insurance and pensions.

Some have pointed out that since the majority of employees in the United States never enjoyed the benefits of the "standard," lifetime, career employment system (e.g., working for one employer over the life of a career) in the first place, the shift to nonstandard work arrangements may not be that great a change for the labor force on average (Cappelli, 1999). Nonstandard work arrangements existed before standard employment relations did; internalized, hierarchical employment relations did not replace market-based ones until the middle part of the 20th century in the United States (see Jacoby, 1985). Furthermore, nonstandard work arrangements never disappeared from the economy, and a peripheral workforce has been used fairly consistently as a buffer to protect the jobs of "permanent" employees (Morse, 1969; Harrison, 1994).

Nonstandard work arrangements may be advantageous for some workers if they enable them to advance their careers. For example, a recent study by the National Association of Staffing Services (1994) found that 78 percent of surveyed workers take temporary jobs in order to get a chance to obtain a full-time job. However, analyses of the 1995 Current Population Survey data suggest that only about 4 percent of men and 6 percent of women who are regular full-time employees previously worked in a nonstandard work arrangement for their current employer. In particular, agency temporaries, on-call workers, part-timers, and some types of contract workers are more likely than regular full-time employees to experience employment instability (change employers, become unemployed, involuntarily drop out of labor force). Nonstandard work may help individuals secure permanent jobs elsewhere by helping them gain work experience or skills, but it does not appear to lead directly to permanent jobs, at least on average. Certainly some nonstandard work also meets the needs of individuals who, for a variety of reasons, prefer the schedules such work provides. Some part of the increase in nonstandard work may reflect a changing preference of employees for such schedules. Independent contractors and other self-employed persons who have nonstandard work, however, do not appear to have high job insecurity or employment instability (Houseman and Polivka, 1998) and perhaps neither need nor are interested in making a transition to a permanent job.

The growth of nonstandard work arrangements represents a change in the work context that has implications for a number of issues discussed in this book. Instead of the employer controlling one's work, for example, persons in some nonstandard work arrangements either control their own work (along the lines of independent contractors and other self-employed persons) or are supervised by someone other than their employer (as are employees of temporary help agencies). Employees of temporary help agencies and contract companies also work in locations other than their employing firm, such as in a client organization. Moreover, some nonstandard work arrangements, such as independent contractors, are characterized by multiple employers. These arrangements represent important challenges to traditional definitions and classifications of employment that rely on the single, long-term employer, including most labor and employment law.

Nonstandard work arrangements may also involve the transfer of control over work from the organization in which a person works to occupational groups (such as those to which independent contractors belong) or to other organizations (such as contract companies). This transfer of control has implications for the object of the worker's attachment (e.g., the occupation rather than the organization) and who should be responsible for training the worker. The importance of occupational analysis—particularly in describing the content of work as performed in a variety of organizations—is therefore increased, since the organization may well become less relevant as the unit within which work is primarily defined and structured.

## Training and Employee Development

Training is an important barometer of changes in work, including those that affect its structure. When the context of work changes significantly, organizations often alter their training to help people adjust. Information trends in training practices can shed light on the changing nature of work. Changes in the topics of training reflect a different competitive environment. For example, a comparison of the *Training Magazine* surveys in 1988 and 1997 shows increasing proportions of organizations offering courses in creativity, strategic planning, and managing change (Lakewood Research, 1988, 1997).

New technologies were the most significant issue affecting training practices, according to a Human Performance Practices survey conducted by the American Society for Training Developers (ASTD). A high proportion of firms responding to the survey offered courses in computers (91 percent), exceeded only by the proportion offering an orientation to the firm (93 percent) and management training (94 percent) (Bassi et al., 1997a). Among companies responding to a survey of workplace skill, 75 percent cited computerization as a factor contributing to needs for additional skills, and 67 percent cited greater teamwork and employee participation practices (Olsten Corporation, 1994). In a study by the Bureau of Labor Statistics (1996a), establishments reported that computer training consumed 20 percent of their training hours—more than were devoted to any other single subject.

The next most important issue affecting training practices among the Human Performance Practices respondents was restructured and redesigned jobs (Bassi et al., 1997). Most establishments provided related training, including technical (88 percent) and quality assurance (76 percent) courses. Among training respondents, the percentage offering technical training rose from 76 percent in 1988 to 91 percent in 1997. Fewer offered quality training—63 percent, up from 50 percent in 1991 (Lakewood Research, 1988, 1991, 1997).

The restructuring trends noted earlier are altering the content of training. The mode of lesson delivery is also changing. Instead of printed sheets handed out in a classroom, electronic delivery provides lessons at the time and place that the worker needs them, a feature that offers learners a significant advantage.

The most recent training survey found that 12 percent of companies used such Electronic Performance Support Systems (EPSS), up from 5 percent in 1996. Among organizations with 10,000 or more employees, 28 percent now use EPSS (Lakewood Research, 1997). Outsourcing has also come to the training function. By one estimate, employers spend nearly equal amounts on internal and external training, although larger organizations spend more on internal training (Bassi et al., 1996).

The concern that people may move more rapidly across jobs, careers, and organizations has helped increase interest in lifelong learning, especially as employees come to believe that they will need to create their own opportunities for education and training. Data that separate employer-required from employee-initiated training are difficult to find.[1] But we do know that the proportion of employed individuals taking one or more courses to improve their current job skills increased from 29.5 percent to 32 percent of all workers between 1991 and 1995 (National Center for Educational Statistics, 1997), a modest trend. The WorkTrends™ data summarized in Table 2.3 also indicate that employees believe that training opportunities have increased moderately for most occupational groupings. And the number of older students in college, most of whom are working or have work experience, has been growing more than the number of younger students, especially among women.

The question of declining job security and stability raised earlier also has important implications for training. In general, firms with higher turnover are less likely to provide employee training. The Bureau of Labor Statistics reported from its 1995 survey that establishments with high labor turnover trained only 7.2 hours per employee compared with 12.5 hours for medium-turnover and 10.8 hours for low-turnover establishments. High-turnover establishments also spent markedly less for in-house and outside trainers and for tuition reimbursements compared with the medium- and low-turnover establishments (1996). This trend was supported by the WorkTrends™ data. Whether surveyed in 1985, 1990, or 1996, workers in organizations that laid off employees

---

[1]The National Household Education Survey acquired both kinds of information but reported them in aggregate.

due to a business downturn during the previous 12 months reported fewer company-provided opportunities to improve their skills (39 percent) compared with those in companies that had not downsized (51 percent). Workers in companies with layoffs also responded less positively than companies with stable employment to the item, "I am satisfied with the opportunities for training and development that my company provides me" (34 versus 47 percent, respectively).

Changes in the structure of work also have implications for practices in addition to training. The first of these is employee selection, broadly defined as how employees are recruited and then selected into the organization. If employee attachment to the employer truly is weakening, resulting in more quits, if firms continue to restructure and churn their workforce, or if they hire more from the outside market, then it is reasonable to conclude, first, that employers will be engaged in more employee selection, and that it will become a more important activity for the organization. The second implication is that the criteria for selection will also change. Skills will become more important, and the potential for development less so, if internal development and career ladders inside the firm give way to more outside hiring. And the type of skills needed will also change. Teamwork will place a premium on teamwork skills and self-motivation and self-management, given the decline in direct supervision. Employers may also find that they rely more on trying to create a corporate culture that effectively substitutes for direct supervision by conveying norms and values that govern performance. If so, then issues of employee "fit" with the culture will become more important.

## The Social/Psychological Contract

The relationship between employer and employee has both formal, written characteristics and informal aspects. It shapes the expectations of employees and their attitudes and behavior toward their work. Virtually all jobs are complex enough that it is impossible to specify in advance all of the duties and performance levels that are required. Consequently, it is difficult to govern them through explicit, legal contracts. The unique problems and situations that inevitably pop up in most workplaces can only be addressed if employees are willing to use their good judgment and improvise solutions that advance the interests of the organization. The social/psychological contract is at the center of efforts to secure the compliance of employees, to get them to act in ways that serve the goals of the employer even when those actions are not in their immediate interests.

When social/psychological contracts are in force, the assumption is that the employment relationship is long term and that the employer has policies that will benefit the employee in the long run. These include income security, in the form of stable careers and lucrative pensions. But the most important of these policies has been promotions, used to reward good performance and to both reward and motivate superior performance. Promotions do not require that employers specify the exact behavior required of employees nor do they demand constant monitoring. The rewards are very desirable in that they typically represent sizable increases in compensation, increases that several studies have found more than compensate for the additional job demands. What persuades workers that they will get the reward of promotion in the long term for good performance is in part that they have seen it happen before.

In addition to these long-term exchanges are psychological mechanisms centered on the notion of reciprocity. The concept of reciprocity refers to the sense of obligation that one feels to repay gifts, a value that has been identified as underlying every culture on the planet (Gouldner, 1960). Studies that follow employees over time find that new entrants believe that they owe their employer a great deal and that the company owes them relatively little, reflecting the sense that they are indebted to their employer. As time goes on, their view of the relationship changes. The longer they are with the company, the more they believe the company owes them (Robinson et al., 1994). This trend may reflect their own investments in the organization. Studies find that, as long as they continue to meet acceptable performance levels, employees tend to get more rights and privileges in organizations the longer they stay (Rousseau and Anton, 1991).

If contracts are voluntary agreements based on promises about the future behavior of the parties, then social/psychological contracts are based on an individual's perception of the appropriate obligations (Rousseau, 1995). Long-standing practices, for example, create expectations about obligations that shape employee behavior over their careers.

Inside most large companies in the United States, the social/psychological contract for exempt (that is, professional and technical) employees has taken a relatively common form. Employers would buffer employees from the vagaries of the market, offering secure jobs and careers and stable growth in compensation associated with predictable promotions and retirement benefits. In return, employees offered acceptable levels of performance and commitment. Commitment took two forms: a willingness to stay with the company, turning away offers of jobs elsewhere, and an acceptance of and identification with the goals of the company.

The contract for nonexempt employees (who must volunteer to work extra hours and be specifically compensated for such work) was surely different. In unionized settings, most of the mutual obligations were set out in elaborate written contracts. The traditional social/psychological contract was probably limited to an expectation that the company would continue to offer long-term jobs (interrupted by periodic but temporary layoffs) and that employees would stay in them. Among progressive nonunion employers, the contract may resemble that for white-collar workers; for less progressive employers, it looks more like a simple market exchange.

The changes in the way work is organized, including downsizings and restructurings, and the move toward nonstandard work seem to have broken the social/psychological contracts described above. When psychological contracts are violated, declines in employee morale are one consequence and others may include an increase in quit rates and declines in performance. The American Management Association found that 72 percent of its surveyed companies that had cut jobs reported an immediate and negative impact on morale, including an increase in absenteeism and disability claims.[2] Employee attitudes also seem to suffer. A

---

[2]Interestingly, all of these effects abated with time. One year later, only 36 percent of these firms reported that morale still suffered (American Management Association, 1996).

recent survey by the Wyatt company observed that morale was substantially lower in companies that had downsized compared with those that had not. The WorkTrends™ survey data support this point, showing that, whether surveyed in 1985, 1990, or 1996, workers in organizations that laid off employees due to a business downturn during the previous 12 months reported less satisfaction with the present employer (49 percent answered satisfied or very satisfied to the item, "Considering everything, how would you rate your overall satisfaction with your company at the present time?") than did those in companies that had not downsized (64 percent answered satisfied or very satisfied). Describing the sharp decline in employee commitment observed over time in his polls, Daniel Yankelovich notes that "Companies are unaware of the dreadful impact they are having. They don't realize they are violating an unwritten but important social contract they have with workers" (*Fortune*, 1994).

According to the WorkTrends™ survey, workers in organizations that laid off employees reported less trust in senior management than did those in companies that had not downsized. In response to the question: "Senior management gives employees a clear picture of the direction the company is headed," 33 percent of respondents whose companies had layoffs agreed or strongly agreed, compared with 47 percent of respondents from companies without layoffs. Furthermore, only 26 percent indicated a belief in the truthfulness of senior management, compared with 38 percent in companies without layoffs.

Employees may also withdraw their cooperation when they perceive that social/psychological contracts have been broken. A series of studies by Sandra Robinson and her colleagues examine perceptions of psychological contracts and their breach (Robinson et al., 1994). Among recent business school graduates hired into their first job after graduation, for example, she found that they still saw their employers as having substantial obligations to them; the employees trusted that those obligations would be honored. And when those obligations were breached, the employees responded by reducing their own obligations and self-reported measures of performance and commitment (Robinson, 1996).

It is difficult, however, to find evidence that employee and company performance declined when companies broke the traditional social/psychological employment contract. To illustrate, the 1994 American Management Association survey on downsizing found that while 86 percent of companies that had downsized reported that employee morale declined, employee productivity rose or held constant in 70 percent of downsized companies; profits rose or held constant in 80 percent (American Management Association). A survey conducted by human resources managers of employers who restructured reported adverse effects on morale and commitment but also increases in productivity, service levels, and greater competence among employees (Wyatt Company, 1993). These reports were based on self-assessments. Despite the enormous changes and the collapse of employee morale, employee performance did not seem to decline enough to affect corporate performance. No doubt the context of downsizing across most of U.S. industry from the 1980s through the mid-1990s caused enough fear among workers to prevent declines in their performance.

Employer efforts to articulate a new deal governing the employment relationship have been driven by an understanding within organizations of the profound way in which they have unilaterally broken the old deal and concerns about the possible negative consequences that doing so may have on employee behavior. Most of these efforts have focused on being explicit about the limits on employer obligations, simply being clear that the employer can no longer guarantee job security and that employees have to look out for themselves. The word "employability" is typically used to describe these employer-initiated efforts to craft a new social/psychological contract. It focuses on the new and much more limited obligation of employers to help (but not provide or ensure) their workers develop skills that will continue to be in demand, if not with their current employer then with another. How employees will respond to this and what else might be part of the emerging social/psychological contract is less clear.

Some indication about the nature of the new employment relationship can be obtained from employee surveys. A survey of 3,300 employees by Towers Perrin asked respondents about their perceptions of the new deal in the workplace. Employees recognized that lifetime employment with one company is unrealistic—and may also be undesirable for them. Only 45 percent felt that they had the opportunity to advance in their current organization, suggesting that the key to advancement in their mind is movement across organizations.

A multiemployer survey asking what employees expect from the new deal (based on employee surveys inside each company) found that "secure employment" is ranked fifth in importance out of 10 attributes that employees expect from their companies. Interesting work, open communications, and opportunities for development took the top three spots (The Conference Board, 1997). Focus groups conducted by Towers Perrin report that while employers put commitment and trust as the two most important things they want from the new deal, employees want professional development and training, investments that make it possible for them to reduce their dependence on their current employer and become more valuable in the labor market (Milligan, 1996).

There is some evidence from the Towers Perrin survey that employees now seem to have rather positive views about their work situation as opposed to earlier in the 1990s: 79 percent believed that their company had treated them fairly and that they had a sense of personal accomplishment from their work, and 75 percent feel motivated and that they could have an impact on company success. The contrast with much more negative survey results just a few years before no doubt reflects several issues: day-to-day experience in the workplace has changed as growth has replaced downsizing; extensive hiring means that a large percentage of the respondents had little experience with either the restructuring period or the old deal; and people tend to recalibrate their perceptions (Bookbinder, 1996).

## CONCLUSIONS AND IMPLICATIONS

The arguments and evidence presented in this chapter suggest several general conclusions about changes in the structure of work. First, there appears to be much greater diversity in how work, jobs, and occupations are structured. The rise of nonstandard work is perhaps the best example, but there is also greater variety in how tasks are performed (e.g., the amount of teamwork and employee involvement). In part because the structure of organizations has become more fluid and the boundary around what tasks are performed inside or outside the firm has blurred, the hierarchy and structure of jobs inside firms and organizations has also blurred. In general, narrow jobs have given way to broader jobs; management positions, especially those in middle management, have been cut, leading to flatter organizational charts and much wider spans of control. Traditional boundaries around jobs, such as the distinction between managerial and production work, white-collar and blue-collar jobs, the barriers around craft work, and the narrow job descriptions of production jobs associated with scientific management have all blurred. Finally, the employment relationship defined as the set of mutual obligations and expectations between employer and employee has substantially weakened. Expectations seem to have clearly moved toward more transient relationships, although the data from the labor market have yet to show substantial changes in that direction.

Central to the theme of this volume are the implications of the changes in work structure for classification systems. For example, if careers increasingly span several employers, occupational structures based on job ladders within companies may become less important and those that cut across organizations may become more so. Career paths may increasingly be built around temporary and part-time jobs, often in combinations that do not necessarily fit with traditional notions of internal mobility. If individuals in fact will be spending more time in similar jobs, then perhaps occupational classification systems will need to broaden considerably in order to accommodate greater depth in a given occupation. For example, the classifications associated with computer programming might need to be broadened to differentiate the range of programming jobs one might have over a lifetime of doing that work and how different positions might build on each other to create an entire career.

As noted earlier, the skills, knowledge, and abilities that are demanded in the changing work structure are also different from those in the past and need to be captured by classification systems. Perhaps the biggest challenges are created by the potential for greater mobility across firms that results in the demand for more lateral hiring. Employers and employees will need better descriptions of jobs and of employee attributes in order to facilitate more external hiring. This information needs to be standard-ized in order to be useful, but the demand for standardization militates against the trend toward broader jobs and more perme-able boundaries between them, a development that leads toward more variations within job titles. Thus, occupational classifica-tion systems may need to allow for this variation and help those designing jobs to better understand the organizational forces that influence how work is done today.

Finally, if, as suggested in this chapter, the scope and pace of organizational restructuring is accelerating, then the speed with which jobs are changing is also increasing. To be accurate, occu-pational classification systems must therefore be updated more frequently. And, as we suggest in Chapter 5, if they are to become a significant aid to decision makers whose actions are shaping work structures, they will need to be transformed from backward-looking tools that describe and classify jobs to more forward-look-ing analytic tools that generate options for how work might be structured in the future.

# 4

# Changes in the Structure and Content of Work

The purpose of this chapter is to explore what workers do and how what they do is changing. The relative growth and decline of broad occupational categories used by the Bureau of Labor Statistics (BLS) are useful for providing a picture of the changing mix of occupations. BLS data clearly show the decline of farm and blue-collar workers and the rise of professional, technical, and what have traditionally been called white-collar (managerial/administrative, marketing/sales) workers. However, they tell us little about what is happening to the nature of work *within* these categories or whether the categories are themselves useful for distinguishing among the activities and experiences of workers in different occupations.

In the sections that follow we examine closely the changes occurring within occupations by using the following broad categories: blue-collar, service, technical/professional, and managerial workers. However, even as we draw on the literature describing the nature of work within each of these conventional categories, we demonstrate that the categories themselves are losing much of their descriptive and analytic meaning. The nature of work is changing not only within these categories but also in ways that blur the traditional distinctions among them. This observation is supported by the WorkTrends™ data reported in

Table 2.3. Statistical regression analyses of worker attitudes on occupational groupings revealed that occupations accounted for a decreasing amount of variance in attitudes toward work and employment between 1985 and 1996. The ability to predict work attitudes based solely on respondents' occupations diminished from 1985 to 1996.[1]

We analyze the nature of work within these categories by considering four dimensions along which work varies and appears to be changing in significant ways:

- The degree of discretion or decision-making power workers have over how to do their jobs. We refer to this as *autonomy-control*.
- The range or breadth of the tasks embedded in a job. We refer to this as *task scope*.
- The substantive (or cognitive) complexity, or the degree of cognitive activity and analysis needed to do a job. We will refer to this dimension as *cognitive complexity*.
- The extent to which the quality of social interactions, including their emotional quality, is critical to job performance. We refer to this as the *relational* or *interactive* dimension of work. It includes *emotional labor*, which is a relatively new concept and an increasingly well-recognized, if not an increasingly important, component of many jobs in which interactions are critical tasks.

We view these as key dimensions of work. In various scientific literatures, they are the primary concepts that have been used to study the relationship between skills and wages, between skills and compensation, and other features of jobs and occupations. The exception is industrial and organizational psychology, which tends to use more refined indicators for skill. These four dimensions also are broadly supported by multivariate studies that factor analyze more detailed measures of work in search of underlying factors or dimensions (for example, see National Research Council, 1980; for a review, see Spenner, 1990). Thus, these

---

[1]A binomial test of this trend (i.e., $R^2$ decreasing versus increasing or remaining constant) was significant ($p < .05$).

dimensions have empirical as well as conceptual merit. Finally, these dimensions open useful conceptual windows on the increasing heterogeneity of work, the debureaucratization of work, increased choices for structuring jobs, and increased interdependence among work structures.

The first two dimensions—autonomy-control and task scope—are well established in the job design and analysis literatures (Hackman and Oldham, 1975; Hackman, 1987). Because autonomy-control reflects the vertical division of authority in an organization, it is also found in the organizational design literature and it parallels the legal distinctions that define employee rights and organizational obligations. Task scope has been the subject of considerable debate over the years and focuses on the horizontal division of labor. Ever since scientific management and early industrial engineers formulated narrow specialization as a principle of job design, scholars and practitioners have debated the trade-offs of specialization versus job enlargement, job rotation, team-based work systems, and other means of expanding the scope of a job.

The cognitive complexity dimension is normally treated in job analysis as the depth of expertise one needs to do a job. In comparison with major sociological approaches, our definition of cognitive complexity is nearly identical to the definition used by Kohn and colleagues in their program of research (Kohn and Slomczynski, 1990). They define substantive (or in our terms cognitive) complexity as the degree to which performance of the work requires thought and independent judgment. Our definition is more restrictive than that proposed by Spenner (1990), who includes not only cognitive demands, but also interpersonal demands and task scope in his definition of substantive complexity.

The social interaction dimension includes both relations between workers and their customers or clients, and relations among workers. Although interpersonal work may long have been important in many jobs, it has become more salient for a number of reasons. Relations between workers and customers have become more prominent because of the growth of customer-contact jobs and because of the increased importance that employers give to competing on the basis of customer service (e.g., Albrecht and Bradford, 1989; Reichheld, 1996; Zemke and Schaff,

1989). Relations between employees at all levels of the organization appear to be more important because of the spread of collaborative forms of work organization. Employers have gone from introducing groups as a stable building block of organizations to using multiple types of permanent and temporary groups to accomplish organizational goals—supervised teams, self-managed teams, cross-functional teams, quality circles, labor-management committees, problem-solving groups, project teams, task forces, top management teams, etc. The job analysis literature has traditionally defined the interactive dimension to include communication and negotiating skills, but has paid little attention to emotional labor. Making someone happy, excited, calm, or committed is a crucial skill in a growing number of jobs.

In the following sections, we also discuss the increasing importance of information technology, not because it represents a fundamentally new dimension of work, but because information technologies are creating an array of new jobs and changing how existing jobs are performed. In addition, we examine the influences of changing markets, changing workforce demographics, changing organizational structures, and changing employment relationships on the structure and content of work.

## BLUE-COLLAR WORK

Blue-collar workers are usually viewed as workers who are nonmanagerial (i.e., covered by the National Labor Relations Act) and nonexempt (i.e., covered by the wage and hour provisions of the Fair Labor Standards Act). Although the Bureau of Labor Statistics does not use "blue-collar work" as a specific occupational category, their categories that come the closest to encompassing the popular notion of blue-collar work are "skilled and semi-skilled production" and "craft workers," "operatives," and "laborers." Using these categories, blue-collar workers represented approximately 25 percent of the labor force in 1996, down from 40 percent in 1950. In this section, we focus on blue-collar work in manufacturing, since this is the typical image associated with this category. Perhaps the most distinguishing feature of blue-collar work is its presumed position in an organization's vertical division of labor. Blue-collar workers are assumed to be supervised by managers and, therefore, to have low levels of autonomy and control over their work. The distinction arises out of the presumption that those who conceive how work is to be done can be separated from those who execute the work. Those who "execute" are, of course, the blue-collar workers. Frederick Taylor's scientific management methods elevated this distinction to a normative principle: conception *should be* separated from execution in order to organize work efficiently and reward workers in ways that satisfy their economic needs. Not surprisingly, this principle has been the subject of debate since it was first enunciated. How much control over work-related decisions is delegated to those doing the work has become an especially important part of job and organizational design decisions, as some firms restructure and move decision making down to lower levels, while others centralize it further with the aid of digital control technologies.

Scientific management also emphasized the scope of blue-collar work by stressing the importance of segmenting work into clearly defined tasks that formed discrete jobs requiring narrow skills. This approach to job design fit well with the growing mass markets and need for standardization associated with the factory system. The design of jobs was perceived largely as an engineering task aimed at producing mainly physical results (the number of boxcars loaded; the number of parts cut, polished, or painted).

## The Changing Nature of Blue-Collar Work

The types of organizational restructuring discussed in Chapter 3 are challenging traditional principles for the design of blue-collar jobs. Blue-collar production work in many firms is expanding to include more decision-making tasks that in the past would have been part of a supervisory or managerial job. This, more than anything else, makes the term "blue-collar work" or "blue-collar workforce" less useful as an analytical or practical tool. Moreover, for some production workers, narrow job definitions are giving way to broader involvement in work teams and interactions with external customers, clients, and patients.

Part of our understanding of the changing nature of blue-collar work builds on research concerning the adoption of "high-involvement" or "high-performance" work systems. The basic argument in this literature is that blue-collar work structures are changing because the highly specialized division of labor that supported cost minimization in mass production is no longer compatible with current markets. Under the logic of mass production, which emphasized quantity over quality, work was divided into individualized, repetitive jobs that required low discretion and skill. Today's markets demand competitiveness on the basis of quality, innovation, and customization (Piore and Sabel, 1984; Appelbaum and Batt, 1994).

Under the logic of high-involvement systems, quality and innovation result from work designed to utilize high skills, discretion, and the participation of frontline workers in operational decision making. Human resource practices such as training, performance-based pay, and employment security provide complementary incentives for participation (Ledford et al., 1992; Osterman, 1994; Kochan and Osterman, 1994). A central hypothesis in this literature is that work structures are part of a larger internal labor market system or set of complementary work and human resource practices (Milgrom and Roberts, 1992). Thus, the content of work must be analyzed as part of this larger system in order to understand how work structures affect outcomes of critical interest to different stakeholders. This point will feature prominently not only in this section on blue-collar work but also in discussion throughout this chapter.

### Effects of Teams on Skills and Work Content

At the heart of the movement to high-performance work systems lies the growing emphasis given to teamwork and work structures organized around work units or groups rather than individual jobs. The decision to implement teams affects both the degree of control delegated to nonsupervisory workers and the scope of tasks that workers are expected to perform. As such, it changes both the vertical and horizontal divisions of labor. Team-based work structures also increase cognitive complexity and the interaction requirements of blue-collar work.

A number of studies demonstrate that both cognitive and interactive skills are becoming more important in blue-collar jobs, as team-based work structures are more widely used in blue-collar work. Morgan et al. (1986) identified giving suggestions/ criticisms, cooperation, communication, building team spirit/morale, adaptability, coordination, and acceptance of suggestions/ criticisms as critical teamwork skills and behaviors. Perhaps the most extensive research on team competencies, however, has been carried out at the Naval Air Warfare Center Training Systems Division (Tannenbaum et al., 1992; Cannon-Bowers et al., 1995). It developed a model of team effectiveness (see Figure 4.1) that focuses on the influences of organizational, work, and task characteristics on requirements for both *individual task competencies* and *team competencies*. Subsequent work has concentrated on decomposing team and task competencies. There are, as seen in Figure 4.2, four basic types of teamwork competencies: "transportable," "task contingent," "team contingent," and "context-driven." The two categories most relevant to occupational

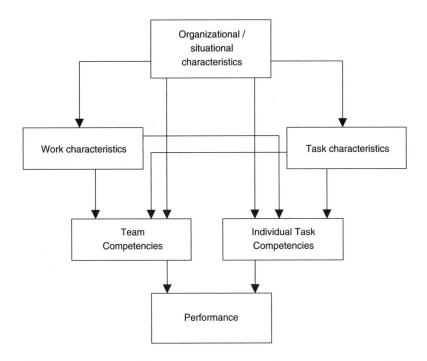

FIGURE 4.1 Team effectiveness model: Conceptual framework. Adapted from Tannenbaum, Beard, and Salas (1992)

|  | Team Generic | Team Specific |
|---|---|---|
| Task Generic | Transportable competencies: generic competencies that generalize to many tasks and team structures.<br><br>Examples: project teams, task forces. Only knowledge really necessary is the knowledge of teamwork skills and behavior. These skills are necessary to individuals involved in multiple teams. | Team-contingent competencies: competencies that are specific to a particular configuration of team members, but not to any particular task situation.<br><br>Task situations include requiring a team to perform competently on a number of different tasks—e.g., quality circles, functional department teams, self-managed work teams. |
| Task Specific | Task-contingent competencies: competencies that are related to a specific task, but that hold across different team member configurations.<br><br>Apply to team situations where membership changes frequently due to organizational policy (learn many jobs) or high personnel turnover (e.g., military cockpit crews who rotate to relieve one another). Knowledge is detailed on the task, not about the teammates. | Context-driven competencies: competencies that are dependent both on a particular task and team configuration. These vary as either the task or team members change.<br><br>Important for teams performing highly demanding tasks— emergency personnel, sports teams, aircrews, military teams.<br><br>Require flexibility and rapid adaptation. |

FIGURE 4.2  Types of team competencies.

analysis are those that are generic across all teams. Thus the focus should be on team competencies that either will generalize across tasks and team situations ("transportable") or that are related to specific tasks regardless of who the other team members are ("task contingent"). Work in these categories requires the development of interactive as well as cognitive skills.

Both qualitative case studies and quantitative studies of the changing nature of blue-collar work have documented the importance of communications, problem solving, and coordination within and across teams (Adler, 1993; MacDuffie, 1996; Rubinstein, forthcoming; Appelbaum and Berg, 1997). Rubinstein found that quality performance in the team-based production process employed at the Saturn Corporation was significantly influenced by the amount of communication among team leaders. His study illustrates the links between cognitive and interactive skills, since he showed how the union at Saturn had created a dense social network among these team leaders that facilitated communication, coordination, and problem solving on production and organizational issues.

Appelbaum and Berg (1999) demonstrated the growing importance of coordination and communication in a three-industry study of work organization. They found that, compared with workers in traditional job structures, those working in teams reported higher levels of participation in problem solving, communications with supervisors and managers both in their units and in other parts of the organization, and greater responsibility for a wide range of duties traditionally reserved for supervisors and managers. One effect of these changes is to blur the traditional lines of demarcation between blue-collar and managerial work in settings that rely heavily on teams, as well as the distinction between blue-collar workers and the technicians and other professionals who design and support the technical and organizational systems associated with their work. Box 4.1 illustrates the blurring of these traditional occupational boundaries in several industries.

Relationship management for blue-collar workers is not limited to shop floor work teams, labor-management committees, or cross-functional problem-solving groups. By reorganizing work from functionally driven lines to product- or customer-focused centers, employers are able to more directly link customers to workers, and blue-collar workers need to develop customer relationship management skills. Box 4.2 provides an example of this; it is taken from a case study of Corning, Inc., a company that has worked collaboratively with the American Flint Glass Workers Union to create team-based work systems.

## Effects of Information Technology

A large number of recent empirical studies have shown that technology is changing blue-collar work in significant but

---

**BOX 4.1**
**Blurring of Occupational Boundaries in the Steel,**
**Apparel, and Electronics Components Industries**
**(Appelbaum and Berg, 1999)**

In many of the plants in the three industries in our sample, workers are expected to engage in problem solving whether or not there are self-directed teams. . . . Front line workers have increased responsibility for coordinating production activities, and offline teams are widespread. Nevertheless, the survey results indicate that self-directed teams have responsibilities that go beyond this, and that there are significant distinctions between these responsibilities and those of other workers. . . . [M]uch of the information in these plants is collected and remains at the bottom levels, where it is acted on directly by workers who call on subject matter experts, confer with workers and managers outside their work groups, and make decisions that affect product quality, maintenance of production equipment, and adherence to production schedules. . . .

In tandem with the reductions in the number of supervisors and with greater reliance on front line workers in many of the plants in our sample, there has been a major change . . . in the supervisors' role. Supervisors are expected to coordinate with purchasing or with earlier stages of the production process about the quality and availability of incoming materials as well as with internal and external customers downstream from them. As we have seen, they participate extensively in offline teams that deal with product quality, cost reduction, equipment purchases or modification, working conditions, or training; and they usually have responsibility for facilitating these meetings. They are also called on to resolve differences of opinion among blue-collar workers and to provide structured on-the-job training for workers, who now require greater skills. Where self-directed teams have been introduced, either formally or de facto, supervisors spend very little time "watching other people work."

---

nondeterministic ways (Hirshhorn, 1984; Jaikumar, 1986; Kelley, 1986; Keefe, 1991; Adler, 1992). Moreover, these studies suggest that past debates over the effects of technology have been framed too narrowly. To ask whether the net effect of technology is to "upskill" or "downskill" blue-collar work fails to capture the

---

**BOX 4.2**
**Teams and Customer Interactions at Corning, Inc.**
**(Batt, 1997)**

Corning, Inc., and the AFGWU [American Flint Glass Workers Union] undertook a joint plant redesign at General Machine Shop [GMS], a supplier of machined parts for Corning's consumer and television tube production plants. The 1992 redesign replaced prior functional departments with product-oriented groups (serving consumer, lighting, and TV products), and within those groups, teams dedicated to producing parts for particular customer plants (e.g., Pressware, Martinsburg, and Greenville plants). The "real change" according to machinists, was ". . . the team-focus on the customer and their product." In the past, customers complained that no one at the plant would even answer their calls. This was due in large part to functional specialization.

Under the team-based system, teams are responsible for end-to-end production—from providing quotes to customers for jobs to arranging for materials from suppliers to interfacing directly with the customer during design and production phases to meeting delivery dates. Because workers are now organized into product-focused teams, customers can directly contact team members on the floor to get updates on the production process and to collaborate over product design and cost. Now, GMS machinists go back and forth on designs and specifications with customers, who also supply the cast iron mold and the blueprint; the machinists also go back and forth with their engineers as needed until they "get it right." The machinists are also more heavily involved in training the workers in customer plants on the new equipment provided by GMS. In addition, each team also has its own annual budget for tools and supplies; whereas in the past, the supervisor had to sign a materials purchase order, machinists now purchase anything under $200. They go to trade shows and interact with equipment suppliers to purchase new equipment. Teams also absorb traditional personnel tasks of scheduling, arranging vacations, and determining overtime.

---

multidimensional ways in which technology affects the content of work.

For example, Zuboff (1989) found that a major effect of information technology on blue-collar work is to replace physical activity with mental and more abstract forms of analysis and response: "Your past physical mobility must be translated into a mental thought process" (p. 71). A worker in Zuboff's study described the implications of this change and its link to the cognitive dimension of work (Zuboff, 1989:73-74):

> Before computers, we didn't have to think as much, just react. You just knew what to do because it was physically there. Now, the most important thing to learn is to think before you do something, to think about what you are planning to do. You have to know which variables are the most critical and therefore what to be most cautious about, what to spend time thinking about before you take action.

The implication of these changes is that information technology changes the mix of skills that are required. "Informated" jobs require less sensory (touch, smell) and less physical skill, and more of what Zuboff labels "intellective skills," such as abstract reasoning, inference, cause-effect analysis, and trust in symbols.

Another important lesson from efforts to introduce advanced technology into work settings has been the need to integrate technology and organizational practices. Shimada and MacDuffie (1987) described the differences between American engineering practices and those found in Japanese auto plants operating in the United States. They observed that U.S. engineers tended to see technology as a standalone technical solution, with work system design considerations to be addressed in the implementation phase of the design and decision-making process. In contrast, Japanese engineers tend to see the hardware or technical features of the work process as inseparable from the human dimensions of the work process.

The differences in effect were brought home vividly in the auto industry in the 1980s by experience of General Motors (GM). In the early 1980s, GM embarked on a major high-technology strategy designed to regain its competitiveness by introducing the most advanced automated technologies available. Over the first half of the 1980s, GM invested over $50 billion in pursuit of this objective, only to find at the end of the investment cycle that it still had the highest-cost manufacturing operations in the industry. The company concluded that it had failed to adequately integrate human resource and work organization considerations with the new technology and subsequently began to revamp its approach to introducing technology.

MacDuffie (1996) went on to document the effects of these two different strategies by showing that high-technology plants were not among the most productive in the industry worldwide. However, using high technology in conjunction with workplace innovations and transformed human resource practices did produce the highest levels of productivity and quality in auto assembly plants around the world. Black and Lynch (1997) also found the percentage of blue-collar employees using computers has a positive effect on productivity of a national sample of manufacturing plants. As in the auto industry studies, the effect of computer use is increased when combined with innovative work practices and cooperative labor-management relations.

## Effects of Industrial/Organizational Restructuring

To illustrate the changing nature of blue-collar work and its consequences, three industry examples are summarized below. Each case illustrates the interdependent nature of the content and contexts of work—that is, the changes occurring in blue-collar work are part of a larger organizational restructuring that encompasses other features of the internal labor market.

*The Steel Industry*    The steel industry in the United States experienced enormous competitive pressures in the 1980s and 1990s that set in motion restructuring and changes in traditional job structures. One detailed case study of the restructuring process in an integrated mill reported reductions in job classifications among skilled trades were reduced from 69 to 16 (Yamagami, 1987). Most integrated mills underwent similar changes and moved toward greater teamwork for both production and craft jobs.

The United Steelworkers of America supported these changes in return for greater voice in organizational decision making at the workplace and, in the 1990s, up through the strategic levels of the firm. The union noted, however, that craft consolidation should proceed without sacrificing the deep knowledge and skills associated with traditional skilled trade jobs: "In the area of trade jobs, the primary objective must be to 'deepen' the skill base of the trades and create tradespeople with greater knowledge within their trade, rather than to 'broaden' the skill base of the trades and create tradespeople with industrial knowledge in a multitude of areas. In some cases, a secondary objective may be to train trades in new skills in order to allow them to complete their job, provided, however, that the integrity of the trade is not undermined" (*USW Guidelines for Participation in Work Organization*, 1992:3, quoted in Klein, 1994:151).

A study of the effects of different degrees of change in job structures and related human resource management practices in this industry demonstrated the value of making these changes. Ichniowski, Shaw, and Prennushi (1997) identified four distinct systems of work organization and complementary human resource practices in this industry. They found that the system embodying the most innovative practices produced the highest levels of productivity and quality.

*The Auto Industry*   The auto industry also underwent major changes in the 1980s and 1990s, again in response to international competitive pressures. Empirical studies (Katz et al., 1987; MacDuffie, 1996) demonstrated that the greatest payoffs to productivity and quality came from a systemic transformation of work structures from traditional narrow jobs to team-based work systems accompanied by changes in complementary human resource and labor-management relations. A traditional auto assembly plant might have over 100 different job titles, whereas at the Saturn Corporation, where work is organized around a team system, there is one title for production workers and six titles for skilled trades. But studies done at Saturn and other parts of the industry (Katz et al., 1987; Rubinstein, forthcoming) have found that the payoffs to quality and productivity depend not only on the team-based structures but also on the extent to which teams are embedded in a broader set of complementary human resource innovations and team members and leaders are engaged in high levels of within-team and cross-team communications and problem solving.

The Japanese-based plants located in the United States— Honda, Toyota, Nissan, Mazda, and Mitsubishi—have all introduced variants on team-based work systems. Partial or fully developed team systems and related work and human resource practices are found in some plants of the large American firms (General Motors, Ford, and Chrysler); however, the majority of these older facilities are characterized by a mix of traditional and team-based systems (MacDuffie and Pil, 1997).

*The Apparel Industry*   The apparel industry epitomizes labor-intensive, low-wage production work. Traditional work organization arrangements in apparel are described as the "bundle system," in which work is divided into narrow individual jobs governed by piece rate compensation systems. Studies of this industry show that this is still the dominant arrangement (Dunlop and Weil, 1996; Appelbaum and Berg, 1999); however, in apparel plants that supply retailers requiring rapid replenishment of stocks (with the aid of shared information systems), the bundle system has been replaced with a team-based modular production system. Modular systems serve to reduce total costs in large part by reducing work-in-process inventories and allowing more flexible, quicker responses to fluctuation in retail sales and other demands. The best estimates are, however, that less than 15 percent of apparel industry work in the United States has shifted to modular systems (Dunlop and Weil, 1996). The rate of change in this industry is therefore dependent on the rate at which retailers require manufacturers to support rapid replenishment strategies.

## Summary

The predominant trends in blue-collar work are toward team-based work systems and toward work that increases the degree of control and task scope and that requires higher cognitive and interactive skills and activities. When advanced technologies are integrated with these changes in work content, team-based work systems achieve higher levels of productivity and quality. Not all blue-collar work is, however, changing in this way: the best estimates are that perhaps one-third of the blue-collar workforce is experiencing these types of changes. Blue-collar work may thus be increasing in diversity at the same time that it is leading to a blurring of the boundaries across blue-collar, managerial, and technical work.

There has been considerable research on blue-collar work and we can summarize with some confidence how this type of work has changed. The autonomy of blue-collar workers has increased to include decisions over process and quality control, formerly the purview of supervisors, in settings in which team organization is practiced. In some cases, new technologies have given blue-collar workers discretion, for example when a data interface rather than a supervisor gives instructions. Worker autonomy, when it occurs, is limited to task autonomy and not strategic autonomy—how to perform work rather than determining the work to be performed.

Not all blue-collar workers have increased discretion, not even all workers organized in teams, as the character of the team makes a difference (Bailyn, 1993). Nor do all workers appreciate increased discretion. Some researchers have found, for example, that work organized in teams may substitute one type of explicit control structure for conformance to more implicit behavioral norms that also limit individual autonomy and discretion (e.g., Barker, 1993).

Job redesign, team structures, and computer-integrated manufacturing have broadened the task scope of much blue-collar work. There has been a clear reduction in the number of job categories and the combining of jobs. For example, workers in self-directed teams often interact with customers and take and track orders, as well as produce the product or service. New technologies have eroded the distinction between some traditional craft boundaries, too. For example, repairs of mechanical equipment may now have electrical and electronic components, with a single person making both types of repairs. The increase in the range of tasks performed by blue-collar workers does not mean, however, that they demand greater skill or increased cognitive complexity.

Craft work has always had high levels of cognitive complexity, but some other blue-collar jobs are increasing in their analytic content. Total quality management programs often include the mastery of productivity modeling, flow charts, and statistical analysis of the production process by the work team, but this varies substantially from setting to setting. More generally, there has been an increase in analytic forms of work at the expense of manual labor.

Finally, we know that the interaction-emotional labor component of blue-collar jobs has increased dramatically. Team-based work and the need to coordinate with customers and technical personnel requires that workers motivate, placate, encourage, and please others as central requirements of their jobs.

## SERVICE WORK

This section considers the changing nature of service work—jobs in which face to face or voice to voice interaction is a fundamental aspect of work (MacDonald and Sirianni, 1996). Service work is historically categorized into three occupational groups: (personal) service, clerical and administrative support, and sales. For simplicity, we use the terms service workers, service jobs, or service occupations in discussing common aspects of these three broad occupational groups. Together these service jobs comprise roughly 41 percent of the workforce. Service jobs grew from 11 to 14 percent of the workforce between 1950 and 1996; clerical jobs grew from 12 to 15 percent; and sales, from 7 to 12 percent. Although women and minorities historically have filled many of these jobs, men have increased their relative numbers. In personal service, male workers rose from 38.3 percent of the workforce in 1983 to 41.9 percent in 1996; in clerical, males rose from 19.4 to 21.2 percent of the workforce; and in sales, from 42.5 to 45.3. This is in contrast to technical and professional work, in which the percentage of men fell by 4 percentage points in technical and 2 percentage points in professional occupations (Current Population Survey merged annual earnings files).

There is considerable variation in the content of work in these occupations along the dimensions we have identified. For example, many clerical occupations circa 1975 were still located in small, local establishments. Business office or customer service staff had moderate levels of autonomy, task scope, and cognitive complexity, more relational interactions with customers, but very low levels of technology use. By contrast, large service organizations included telephone operators, data processors, and typing pools, all of whom had very low levels of autonomy, task scope, and cognitive complexity; their interactions were transactional in nature, and their jobs were heavily mediated by technology and automation. Secretaries could score low on all five dimensions of job content if they worked in a typing pool or moderate to high if they were the only secretary in a small business or if they served an executive who delegated considerable authority and discretion to his or her assistant.

In personal services, there would have been similar variation. The job content of restaurant work would have ranged from very low on all dimensions of work in a fast food restaurant to moderate at a local diner, and relatively high if the waiter was employed at a five-star restaurant. Most health care and child care workers and domestic help would score low on cognitive complexity and technology use, but moderate on autonomy and task variety (because their jobs are not highly rationalized) and high on relational interaction, when the affective or emotional content of jobs is very demanding.

In sales, the content of work would have varied both by product complexity and by customer or client. Given the fact that manufacturing was characterized by the mass production of standardized goods, however, product variety and complexity was limited, and this in turn led to considerable standardization in the content of sales jobs. Retail sales workers in large department stores would have scored low on all five dimensions of work, but sales workers in small businesses would have scored higher. Jobs that involved the sale of more complex or luxury goods (industrial products, wholesale distributors, yachts) encompassed considerably higher levels of autonomy, variety, complexity, and relational interactive content. In summary, although many of the jobs in service occupations required low formal education and received low pay, the range of variation in work along other dimensions was considerable, varying by occupational specialty and organizational context.

The common denominator in all service work is the interaction with a customer—what researchers have referred to as "interactive service work" (Leidner, 1993:26-27). Leidner argues that there are three types of interactive service work in which the success of the work depends on the quality of the interaction. In one type, the interaction is inseparable from the product being sold or delivered, for instance, in child care. In the second type, a product exists apart from the interaction, but a particular type of experience is an important part of the service. For example, airline passengers who buy tickets primarily to get from one place to another are promised friendly service on their journey (Hochschild, 1983). Finally, in some jobs the interaction is a crucial part of the work process even though it is not part of a product being sold or provided. The success of salespeople, fundraisers, bill collectors, and survey interviewers depends on the workers' ability to construct particular kinds of interactions.

The content of interactive service work differs in fundamental ways from the content of production work in manufacturing for several reasons. First, services are produced and consumed simultaneously, so there is little or no opportunity for post-production quality control. Second, the customer typically participates in production. The consumer as "coproducer" is extremely important in defining work content, both because of variation in customer characteristics and demands and because the customer introduces uncertainty into the service production process. As discussed below, control over customer uncertainty plays an important role in management strategies—perhaps as important as control over work processes and behaviors of employees. Third, services are intangible, heterogeneous (no two alike) and perishable (they cannot be inventoried), making performance measurement systems more difficult to develop and implement (Bowen and Schneider, 1988; Zimmerman and Enell, 1988).

In general, interactive service work can be categorized as "relational" or as "rationalized" (Herzenberg et al., 1998; Gutek, 1995; Levitt, 1972). Relational service work emphasizes the personal relationship between the service provider and the customer. Over time the content of such work emerges through the interaction of customer and provider and is defined in such terms as how attachments, trust, and interdependence are built into the relationships. Examples of relational service work include personal service workers (gardeners, hairdressers, housekeepers) and some local service providers (bank tellers, cashiers at the corner store). Such jobs are generally characterized by significant autonomy and variety, moderate cognitive complexity, and minimal use of technology.

The rationalized approach to service work, in contrast, emphasizes routinization. It adopts standardized rules of behavior (Leidner, 1993) and a production-line approach to delivery of services, entailing division of labor, mechanization, and management of customer behavior to achieve conformance with the operation of the system (Lovelock, 1990). Examples of this type of service work are the scripted behaviors of flight attendants (Hochschild, 1983) and the routinized activities of telephone operators (Norwood, 1990; Kohl, 1993) and clerical workers (Garson, 1975; Lowe, 1987; Anderson, 1988; Fine, 1990). Rationalized service work typically reduces autonomy, task scope, and cognitive complexity and can be associated with significant use of technology.

It is important to note that although many of the jobs in service occupations have required low formal education and received low pay, and generalizations can be made with respect to the characteristics of relational and rationalized service work, the variation in work processes is considerable. These variations in autonomy, task scope, cognitive complexity, and use of technology run across occupational specialties and organizational contexts.

## The Changing Nature of Service Work

In the last two decades, the direction of change in service occupations appears to be twofold. First, there has been a blurring of boundaries across these occupations. Clerical and sales jobs, and to a lesser extent personal service, have come to resemble each other more closely because of the further diffusion of work rules and advanced information systems—a process that sociologists refer to as *rationalization*. Modern management strategies have served to integrate service and sales as well. Second, there has been an increase in variation within the occupational categories due to changes in markets (the proliferation of customized goods and services) and to variation in customer service strategies (ranging from highly transactional to relationship-oriented). For clerical and sales work, both changes are arguably different in kind and in pace than changes in the past. Personal service work, by contrast, is much more stable in content, and rationalization of work behavior has continued to take place in ways that are consistent with the past.

## Common Strategies and Technologies Across Occupations

Advances in computer technologies and information systems have facilitated the further diffusion of mechanization and automation in service work. A recent management text notes, "Led by franchisers, more and more service firms are standardizing their operating procedures. Costs are reduced as a result of economies of scale, and bottlenecks become easier to identify and eliminate. Quality control is aided by increased conformance to clear specifications. And standardization of job tasks allows the organization to recruit relatively unskilled, inexpensive workers who require only limited training to perform highly routinized tasks" (Lovelock, 1990:352). Evidence of the expansion of this kind of rationalization to more occupational subgroups is found in a series of studies of clerical and retail workers (e.g., Bluestone et al., 1981; National Research Council, 1986b; Garson, 1988; Kohl, 1993; Batt and Keefe, 1999). Electronic monitoring, once associated only with telephone operator jobs, is now available for use in covering a much broader range of work.

More recently, automation has expanded from the back office (typists, data processors, operators) to the front office (customer service and sales employees). Mechanisms such as toll-free telephone numbers, automatic call distribution and routing systems, and voice recognition systems have made it possible to achieve dramatic improvements in economies of scale through centralized distribution channels serving wider geographic areas. What were once local customer service and/or sales operations providing personal service to repeat customers have become large "call centers" providing remote service through toll-free numbers. Examples include call centers in telemarketing operations, banking (Hunter, 1998a), telecommunications (Batt and Keefe, 1999), and insurance (Keltner and Jenson, 1998). Although research on information technology in the 1980s found no increase in productivity associated with investments in technology (e.g., National Research Council, 1994), more recent studies are beginning to change that view (e.g., Brynjolfsson and Yang, 1996). Automation

of low-skilled and redundant clerical work has led to productivity gains. The remaining jobs may require more skill but may also
incorporate additional routine work tasks (see, for example, Levy
and Murnane's 1996 study of accounting jobs in banking).

The shift from small local offices to large-scale mega-centers
changes the content of work in a number of ways: (a) it increases
the standardization of work (reducing autonomy, task scope); (b)
it increases technology mediation; (c) it shifts knowledge from
specific, tacit, and substantive knowledge of customers to formal
knowledge of programs and procedures; and (d) it shifts social
interactions from relational to transactional. For example, a large
northeastern telephone company consolidated several local offices of 75 employees each into a regional mega-center of 600
workers. According to one account (Mary Batt, 1999, personal
communication):

> In the old office there were 70 or 80 of us. . . . We knew the crews in
> the area, and could call them to find out where an installation stood.
> We knew where the cables were down because of weather prob-
> lems . . . everyone knew each other . . . we used to talk to each other
> about problems . . . there were more informal arrangements for get-
> ting things done. Now, there's reams of paper, too much to read,
> and new product information that comes online. If there's a service
> problem we can't handle, we're supposed to send a note to special
> reps, but don't go and talk to them ourselves. Now we don't have
> to leave our desk for anything. . . . Now with 600 [workers], stan-
> dardization is the rule. . . . We're supposed to adhere to our sched-
> ule [be on-line taking calls] 85 percent of the day. Calls are 340
> seconds each. If I'm talking to a customer and it's ten to twelve and
> I'm on until twenty after, the system cannot be automatically ad-
> justed so that I'm not penalized if I was supposed to go to lunch at
> noon. Instead I'm out of adherence and I have to call into the mana-
> gerial force.

Recent examples of the use of work rules to routinize or script
customer/provider interactions include Leidner's ethnography
of MacDonalds workers (1993), Biggart's (1988) study of direct
sales workers, and Butterfield's (1985) study of Amway. They
show how service work rules are more all-encompassing and in-
vasive than in the past: "quality control is not a matter of standardizing products but of standardizing the workers themselves.
This involves extending organizational control to aspects of the

workers' selves that are usually considered matters of personal choice or judgment" (Leidner, 1993:24-25).

At the same time, however, and unlike in the past, there is evidence of an opposite trend—an overall increase in technical skill requirements and cognitive complexity of service jobs. The initial impact of information technology involved a shift from manual to computer-mediated information processing, but more recent applications involve the manipulation of a variety of software programs and databases. Even more recently, the rapid diffusion of access to the Internet has increased the potential for greater information processing and cognitive complexity. Secretaries at universities, for example, now often act more like research assistants than typists or receptionists. In the mid-1980s, the National Research Council (1986b) signaled the potential for information technologies to increase the skill and complexity of clerical work; over the last decade, there is more evidence of that trend occurring.

An additional source of cognitive complexity is the growth of business strategies that compete on product variety, customization, and innovation rather than low-cost, standardized goods. Workers who service and sell "mass customized" goods with shorter product development time and shorter product life cycles (Pine, 1993) must absorb much more product knowledge and constantly changing information that corresponds to the particular features, pricing, servicing, and legal regulations governing the products for which they are responsible.

Many companies also have adopted business strategies that compete on service quality, as advocated by management consultants from the mid-1980s on (e.g., George and Marshall, 1985; Albrecht and Bradford, 1989; Zemke and Schaff, 1989; Heskett et al., 1990; Schneider and Bowen, 1995). Arguably, to do so requires the redesign of customer service approaches so that workers have more autonomy, a variety of responses, and the ability to interact personally with customers and provide "one-stop shopping" (Schlesinger and Heskett, 1991). The approach advocates a "bridge from service to sales"—blurring the line between occupations that primarily service the customer (inquiries, billing, repairs) and those that primarily sell.

In some ways, these trends represent an attempt to return to the type of personalized or relational approach to services of the past. The difference is that service providers are now based in large-scale organizations and service is customized through the use of databases of information. The customer usually develops a relationship with the organization (through brand loyalty, frequent flyer programs, special cards, etc.) rather than a particular employee—what Gutek calls "pseudo-relationships" (1995:197-211). In any case, the content of service work involves considerably more discretion, variety, and cognitive complexity and use of technology than that found in a strict production line approach. Case studies of the adoption of this approach are found in a wide range of personal service, clerical, and sales work across various industries—including nursing homes (Eaton, 1997; Hunter, 1998b), insurance (Carre, 1997; 9to5, 1992), secretaries in manufacturing concerns (9to5, 1992), retail banking (Hunter and Lafkas, 1998), commercial banking (Keltner and Finegold, 1996), telecommunications (Batt, 1999a), and retail stores such as Home Depot and 9 West (Bailey and Bernhardt, 1997). Examples also exist of better performance outcomes associated with the use of self-managed teams in services (see Cohen and Bailey, 1997). In sum, there is evidence of adoption of both strategies of rationalization and of relationship management in the last two decades.

### Differentiation and Stratification Within Occupations

The second dynamic of change—variation in the skill content of jobs within occupations—has also occurred to a greater extent in clerical and sales work than in personal services. Within-occupation variation has grown as a result of the combined effects of new technologies and markets and management strategies in response to these changes. Increased variation in products and services has led to a growth of specialization in service jobs not unlike what has occurred in technical and professional occupations. In addition to product-based specialization, customer segmentation strategies have created new forms of within-occupation stratification that differ in important ways from the past. One indicator of this change is in the extent of variation in the level of formal education within occupations. Notably, between 1983 and 1996, variation decreased in all occupations (as older individuals retire) except sales, in which it increased slightly.

With respect to product-based specialization, the argument is that, under mass production, new technologies gave rise to new generations of products and services, but variation within each product line was not great. Under mass customization, if workers are responsible for too many different products, the costs of training increase and the ability to sell or provide quality service decreases. For both quality and efficiency reasons, therefore, there is a tendency to shift to greater specialization in jobs in service delivery and distribution. This increased specialization does not show up in statistical data, however; a sales representative is still referred to as a sales representative, although the qualitative evidence suggests that specialization within the category has grown.

Differentiation by customer segment, made possible through advances in marketing research and information systems, has also emerged as an organizing principle for work (Ames and Hlavacek, 1989; Day, 1990; Whitely, 1991). Moreover, variation in the ability to pay has increased with the growth of income inequality, with the population at the higher end of the distribution with more disposable income for luxury goods and services. A succinct statement of the principle comes from Pine (1993:223): "The basics of market-driven management are to segment, target, position, and create. *Segment* your customers and potential customers into meaningful groups that have homogeneous needs within each group. *Target* those market segments that (1) match the capabilities of the firm and (2) have the highest business potential (generally done in terms of revenue, profit, or return on assets). *Position* your firm and its existing and potential products and services in each of the target segments; positioning provides the reason for being, the unique differentiating characteristics that would cause targeted customers to purchase from you. Finally, *create* the products and services that meet the requirements of your target market segments."

Customer segments are then linked to groups of service providers, with variation in the content of work and human resource practices driven by customer characteristics. There is nothing new about attempts to find the right match between the social characteristics of the service workforce and the clients that they serve.   An early example is AT&T's historic selection of white middle-class, highly educated women to fill office jobs so that the public face of the telephone company would match the characteristics of the middle- and upper-class people who could afford telephones. After passing extensive entrance examinations, women received intensive training in rules of behavior, speaking, and scripts that constituted an important dimension of work content (Schacht, 1985).  The large sociological literature on stratification also shows the historic significance of gender, race, age, and sexuality as important characteristics in defining the content of work (Hochschild, 1983; Rollins, 1985; Woody, 1989).

What is qualitatively different is that new matching processes are based not on identity characteristics per se, but on skill, specialized knowledge, or formal educational criteria.  In remote service centers, automatic call distribution and skill-based routing systems link customer segments directly to service workers based on specialized customer or product knowledge, language, or other characteristics. The result has been a distinct reconfiguration of customer service and sales jobs:  these jobs are more differentiated by skill and education requirements, discretion, use of technology, compensation, and career opportunities.  Service jobs at the low end are transactional and have low autonomy, variety, and complexity but high levels of computer-related skills. Jobs at the high end are the opposite.  This approach to work organization is in distinct contrast to the direction of change in blue-collar jobs in manufacturing, in which high-involvement work systems offering high skill and high pay, such as those at Toyota and Saturn, produce cars for the large low-end or middle-market customer base.

An example of matching based on product complexity comes from an information services help desk.  Callers who phone in to ask for help are sorted according to the type of software program they need help on, for example Microsoft Word.  A skill-based routing system links them to a worker who specializes in providing support for the program.  A skill-based pay and occupation ladder is in place so that, once certified in a simple program, an employee can be trained and certified on the next more complex program at a higher pay grade, provided there is a need.  There are 12 grades in all; however, no employee is allowed to provide support for more than 3 programs at once because customer service tends to suffer if employees generalize. Interactions with customers are much more transactional for calls pertaining to simple programs, but they increase in relational content as the complexity of the software and the problems increase.

An example of customer segmentation comes from telecommunications (Batt and Keefe, 1999). In the past, business office employees had broadly defined jobs, handled a wide variety of customers and types of inquiries, and received the same rate of pay. They provided personal, face-to-face and phone service to an undifferentiated public. Under reorganized market-driven business units, they now serve particular customer segments: residential (retail), small business, and large businesses or institutions. Job functions are further divided into sales and service, billing, collections, and repair services.

A typical *residential* call center houses between 500 and 1,500 customer service representatives, who handle 90-100 customers per day and have a call cycle time of about 3-5 minutes. They complete transactions with customers on-line and are discouraged from interacting with fellow employees. As soon as one call ends, an automatic call distribution automatically sends another customer call to the "open" representative. Despite the use of automation and reductions in cycle time, these jobs are not low-skilled; they usually require manipulation of several software systems, knowledge of the product and service information in databases, the ability to integrate new product information on a daily basis due to constant change in marketing campaigns, and customer management and negotiation skills. The extent to which customer interactions are scripted varies across employers. Customer service representatives need to master the art of sincere and authentic customer interactions while manipulating several databases or reading from a scripted text. Electronic monitoring records the content of customer-employee interactions and the time employees spend in each type of work activity.

Call centers for *small business* representatives generally house 100 to 200 employees, and business representatives handle approximately 30 calls per day. Because their orders are somewhat complex, they cannot be handled on-line with the customer on hold. Rather, the business representative takes down the information, enters some of it into a computerized database, but spends considerable time off the telephone completing the order. More complex order entry may also be handed off to another employee. The pace in small business centers is moderate, and representatives freely consult with each other to solve nonroutine problems or get advice on how to handle a customer.

For *large business and institutional accounts*, companies hire college-educated account executives who are supposed who serve as case managers to corporate clients. Job titles vary by the value of the client served. Employees provide customized and personalized service through on-site and electronic exchanges and usually rely on teams of additional support staff and customer service representatives to handle the mechanics of orders and service requests. Advanced information systems serve as a resource for product information and service order processing, but it is not the main tool that case managers or account executives use as they spend considerable portions of their day in the field. Skill in the manipulation of computer software and databases is not as important here as it is for customer service representatives serving customers with less complex needs.

Evidence of the extent of diffusion of segmentation strategies is scarce. Keltner and Jenson's (1998) quantitative case study of banks, insurance, and telecommunications shows that segmentation strategies are viable in each, but do not address the diffusion of segmentation strategies throughout the industries. Batt (1999b) finds that customer segmentation is a primary determinant of work content in a nationally representative sample of customer service and sales establishments in telecommunications. Hunter (1998a), by contrast, finds that retail banks have had difficulty developing a coherent strategy of customer segmentation because segments are difficult to identify and because top managers disagree on the future functions of branch banks.

## Summary

Existing research suggests that, in jobs serving the mass market, there has been a shift from unsystematic and labor-intensive work to more routinized processes through consolidation into remote delivery channels, adoption of standard work practices, and the use of information systems. At the same time, there has been some increase in specialization, information processing, and cognitive complexity associated with mass customization. When these trends converge, the content of work in these jobs includes less independent discretion and scope than in the past, more surface acting than involvement in relationships or negotiation with customers, and more computer use and cognitive complexity. Whereas standardization or rationalization of work has been an ongoing historical process, the combination of process reengineering, information systems, and customization has produced jobs that differ considerably in content from those of the past. These changes have cut across jobs in personal, clerical, and sales occupations.

At the same time, variation within occupations is greater than in the past because of greater variation in product demand and customer characteristics. The majority of jobs continue to serve the mass market, but jobs serving high-end retail, specialized niches, and business clients offer greater opportunities for independent discretion and complexity, and require a more sophisticated set of communication and negotiating skills than previously. However, given the ubiquitousness of computer-mediated work, a much larger proportion of service work, even in these more autonomous settings, is electronically monitored and measured by standard performance criteria.

This review of historically low-skilled occupations has some interesting implications for the military in terms of opportunities for reentry into civilian life. Given the type of technical training that the military provides and the predominantly male recruits, the low-skilled, predominantly female service occupations are not of particular interest. There may be a better match of skills, however, in information service jobs that offer greater cognitive complexity and opportunities for the use of computer skills. To the extent that rationalization and other management strategies improve productivity and added value, these jobs may also offer higher wages. In addition, the relative proportion of men in service jobs has increased, and to the extent that services are provided remotely rather than in person, the importance of gender in employee selection is likely to decline.

## MANAGERIAL WORK

Managerial workers, a broad and diverse group, generally include managers, executives, and administrators. One way of thinking about what managers do is to argue that they process information and pass on the results to higher levels (Scott, 1981:215). Technical developments that increase the effectiveness with which information can be processed may therefore reduce demand for managers. Clearly the explosion of distributed computer power in recent years lends considerable plausibility to this explanation.

It does not necessarily follow that technical change that can substitute for managers will reduce their overall employment. This is because the change may be so powerful that product price reductions will increase product demand sufficiently to offset the reduced use of managers for a fixed level of production. It is also true that some managers—for example, managers of information systems—are complements to computers. Nonetheless, what evidence we have suggests that increased computer power will lead to a fall in managerial employment (e.g., Osterman, 1996). Scott, O'Shaughnessy, and Cappelli describe how the spread of expert systems in the insurance industry has reduced the need for some categories of managers (Scott et al., 1996).

Greater efficiencies can be gained not simply through physical technology, such as computers, but also via new ideas and insights about how to organize people. There is good reason to believe that this is an important part of the recent story.

### Effects of Organizational and Internal Labor Market Restructuring

In the decades from the end of World War II to the mid-1970s, the role of management was viewed in both the academic and the business literature as a defense and justification of formal and fairly lengthy managerial bureaucracies (Chandler, 1977; Thompson, 1967; Williamson et al., 1975). The organizational forms implied by this research were widely accepted when the environment was stable. From the mid-1970s onward, however, the economic environment facing American firms became increasingly unpredictable, and this turbulence brought with it new ideas about how to manage organizations, many of which have been inspired by Japanese management practices. Production processes, both in manufacturing and in services, are being transformed via innovations such as total quality management and just-in-time production. As we discussed in Chapter 3, management itself is changing, as ad hoc teams become more common and firms seek to break down traditional internal boundaries, such as those between design and manufacturing or between marketing and manufacturing.

Changes in the way firms are organized have exerted changes on the career structure of managers; however, it is also apparent that the trends are not unidirectional. Indeed, two quite different tendencies that appear in the data can be characterized as a conflict between cost-cutting and empowerment (Batt, 1996). Another useful distinction is between centralization and decentralization.

In the cost-cutting/centralization model, local autonomy is reduced and layers of managers eliminated. This is made possible by concentrating more power in headquarters and making greater use of information technology and expert systems to manage. This model essentially represents an effort to perfect the traditional hierarchical model of organizations. It does not represent a fundamental shift in views about the purpose or structure of organizational forms. By contrast, the empowerment/decentralization model is based on a different vision of the firm, one in which organizational boundaries become diffuse (due to networks across organizations and teams within them) and in which managers have greater responsibility and discretion for managing these relations.

## The Changing Nature of Managerial Work

Traditionally, the most distinguishing feature of managerial work has been its place in the vertical division of labor. From the earliest days of classical management theory and scientific management, the function of management was conceived of as planning, directing, supervising, and controlling the use of resources to achieve the organization's goals. Although these functions and responsibilities have not changed significantly, the ways in which they are carried out and the organizational and employment contexts and conditions under which managers work are changing as American firms restructure what they do and how they work. With management positions normally came status, power over subordinates and resources, and employment security. However, as corporations have downsized, flattened their large bureaucracies, and moved from impersonal and authoritative controls to team-based peer controls, the role of managers has changed. New forms of work organization are based less on individuals' compliance with rules or supervisors' orders, and more on their commitment to organizational goals, fellow workers, and to satisfying their customers both inside and outside the organization.

In these new organizational forms, managers' jobs have been redefined. No longer supervisors issuing commands, they are now called on to be social supports and coaches of the teams that nominally work for them. Donnelly and Kezsbom (1994:33-41) describe the transition of many organizations to a structure based on horizontal, interdisciplinary project teams. The researchers surveyed 238 project specialists, project managers, and their functional managers, in a variety of technology-oriented companies, to ascertain which skills and competencies are required to effectively lead cross-functional, multidisciplinary teams. According to the results of their survey, five fundamental skills are necessary in the new organization:

1. *Managerial competency*, the interest and ability to use varying combinations of directive and supportive relationships appropriate to the particular subordinate and situation;

2. *Analytical competency*, the ability to understand complex situations, make decisions involving many elements, and retain overall vision;

3. *Integrative competency*, the ability to synthesize a useful systemic outcome from the varying points of view expressed by individuals and organizational units;

4. *Collaborative competency*, interest and ability to work with others, placing value on the formation of alliances; and

5. *Organizational know-how*, the ability to practice one's work within the context of the organization, insight into organizational

culture and politics, as well as the effective use of political influence.

The results of this survey clearly show that managerial jobs in project team organizations involve the successful management of social processes within the team, as well as relations between teams and the rest of the organization.

Another trend is the growing importance of skills in dealing with organizations and people external to the firm. For senior management, according to Useem (1996), this means a focus on external investors and relationships with groups such as boards of directors. For manufacturing managers, there is an increased premium on the capacity to work effectively with external suppliers as the boundary of the firm has weakened and more operations take place through outsourcing and joint ventures (Cappelli, 1999). The importance of teams is growing, as is the consequent premium on team skills. As discussed in Chapter 3, required skills are becoming less functionally oriented and, in some senses, broader in response to the demands of managerial teams that substitute for bureaucratic hierarchies. Managers need to be much more flexible, working in organizations with multiple bosses and at the same time having many more direct reporting channels as the hierarchies flatten.

An important aspect of managers' jobs today is the requirement that they "coach," and many business press and managerial articles have explored this metaphor. For example, Hodes (1992) compares business to sports, with the manager as coach. According to him, in sports a game is fun, an activity that engages one passionately and voluntarily. All players, regardless of their skill level, need coaches who are instrumental to their development. Coaches never get on the field because their main task is to ensure the effectiveness of the players. In the coaching model, Hodes says, the manager is a resource for employees and his or her role is to empower them. Standing on the sidelines gives the coach a perspective different from those on the field and allows him or her to intervene in the process, not in the execution of the work. Critical to coaching in sports and business is maintaining the emotional energy and direction of the team. Emotion management by supervisors is often described as "leadership," "facilitation," and "team support." Like sports team coaches, modern managers' repertoire of skills includes inspiring commitment and high standards of performance, encouraging passion for the product and service to the customer, and facilitating relations between workers (Manz et al., 1990; Hutzel and Varney, 1992; Jackson and Humble, 1994).

In industries in which innovation, rather than standardized performance, is at a premium, managers-as-coaches have become a common conceptualization of the managerial role. In contrast, a traditional, hierarchical structure assumes that "the manager knows best" and does not provide for employees' input. Olalla and Echeverria (1996:16) argue that, when managers serve as coaches, they can unlock knowledge held by team members and facilitate interactions that lead to innovation. The job of coach is to "provide resources, remove obstacles and support the team's well-being so that it can learn, solve problems and continually enhance its effectiveness." However, team-based work can also lead to conflict, emotional exhaustion, and misplaced euphoria, demanding the emotional management skills of facilitators or coaches. Olalla and Echeverria, typical of the managerial literature, argue that managers should have a wide range of emotion management and social process skills and should be able to facilitate open conversations that support productivity, display sensitivity to employees' moods, generate trust, help employees accept their shortcomings so they can learn and move on, listen and observe in order to identify breakdowns in conversational flow, promote autonomy, ask rather than tell, and accept the emotional aspect of work rather than promote a "rational" orientation. As is typical of the literature, the authors base their list on nonsystematic observation. However, if this list is indeed representative of work in "new" organizations, there has been an incorporation of multiple forms of emotional labor into managerial jobs.

Confirmation of these general statements comes from case studies of managerial work. In her research at a Regional Bell Operating Company, Rosemary Batt conducted surveys of managers. She found that (1996:69):

> The overall picture that emerges from survey data is of managers
> in the midst of a transition to a more decentralized and participa-

tory culture along some dimensions of work, but constrained and frustrated by top management decisions with respect of cost and downsizing. . . . Over 60% of lower-level managers in customer services cited technical (computer skills) as the most important new ones, whereas 75% of middle mangers cited "soft" skills in leadership, general management, quality, and labor relations. . . . With respect to decentralization of decision making, the evidence shows that middle and lower managers are experiencing more discretion. . . . The evolution of a more participatory culture is also evident: three-quarters of managers surveyed had participated in at least one form of collaborative or problem solving team.

Similar patterns emerge in other industries. For example, John Paul MacDuffie examined the changing role of managers in automobile companies with the introduction of "lean production," the system of operation that all American firms are borrowing from the Japanese. He found that (1996:106):

> Four . . . observations about lean production lie at the heart of changes in conceptions of managerial authority and responsibility: (1) Managers are much more exposed to market pressures and customer demands under lean production . . . ; (2) Managers are no longer able to devote themselves exclusively to "conceptual" tasks given the absence of protective buffers that shield them from "execution" tasks, yet as a result they develop a greater breadth of knowledge about the production system to guide their subordinates in integrative problem solving; (3) managers are more concerned about the process of decision making undertaken by their subordinates and are less likely to make substantive decisions by fiat . . . (4) managers are forced into much greater independence with managers from other functions and even from other companies.

Finally, Sara Beckman describes similar patterns at the high-technology firm, Hewlett-Packard (1996:167-173):

> That environment [for managers] is more networked externally, with more critical vendor relationships and tighter outbound partnerships, as well as internally across both functional and divisional boundaries. . . . Many manufacturing managers in today's environment share the need to accomplish their objectives through influence rather than direct control. Forced to operate in teams . . . interpersonal, negotiation, and business skills are far more critical than technical skills.

## Summary

In the post-World War II era, there were several studies of the rise of the new class of managerial workers. Books such as *White-Collar* (Mills, 1951) and *The Organization Man* (Whyte, 1956) chronicled the personal and professional lives of the corps of people, mostly men, managing the large new firms of America's industrial economy. In recent years, however, there has been little scholarly study of this group of workers, and we can say less with certainty about how their jobs have changed than with any other group of workers. Much of what we think we know comes from the popular press and by inference from popular management books.

There appears to be some loss of directive autonomy by managerial workers, particularly loss of control over subordinates by low-level supervisors. Many of these managers are now coaches and facilitators of work teams. The rhetoric of management has changed to conceptualize supervisory work as responsible for mentoring and supporting subordinates.

The scope of managerial work appears to have moved away from concern with downward control issues to a focus outward toward colleagues in the organization and toward customers. Because more work in firms is organized by project, managers are increasingly project managers rather than functional managers. Project managers manage the process and flow of work, rather than people.

Traditional management work was cognitively demanding, but new ways of managing are demanding different forms of cognition. Managers now need coordination skills as much as control skills, and many managers appear to have direct responsibility for the financial performance of their units or projects. Indeed, performance-based pay, such as stock options, appears to be increasingly important for all levels of management, not just executives.

Managers no longer appear to have job security. Like blue-collar workers, more of them are subject to dismissal when business declines. This, along with a speed-up of work generally, probably has made managerial labor more subject to stress.

## PROFESSIONAL AND TECHNICAL WORK

The most significant development in professional and technical work is its increase in number and importance in the overall workforce and economy. The professional and technical workforce has figured prominently in discussions of the future of work since Daniel Bell (1973) published his influential treatise on the rise of the service economy 25 years ago. Bell was among the first to argue that experts and knowledge workers would become increasingly critical to economic growth in industrialized countries. Although he conceded that managers and lower-level service workers might constitute the majority of a postindustrial workforce, Bell argued that its elite would be those who create and deploy scientific and technical knowledge. Among the elite Bell counted scientists, engineers, doctors, lawyers, computer programmers, technicians, and most other occupations that sociologists and economists classify as professions or semiprofessions. Although Bell's thesis was hotly contested until the late 1980s, trends seem to have vindicated his vision (Block, 1990; Barley, 1996a).

Analysts most commonly define the professional and technical labor force using the two broad occupational categories of the Bureau of Labor Statistics, "Professional Specialty Occupations" and "Technicians and Related Support Personnel."[2] The occupations that fall within these categories are listed in Box 4.3. Since mid-century these occupations and the work they perform have become increasingly prominent in the U.S. economy, growing 125 percent to become 18 percent of the workforce in 1996. Although there is a long-standing debate among sociologists over how to define who is and who is not a professional (Wilensky, 1964; Carey and Eck, 1984; Nelson and Barley, 1993), we use these categories as our working definition for the purposes of this analysis. We

---

[2]We combine the Bureau of Labor Statistics categories "Professional Specialties" and "Technicians and Related Support Occupations" in discussing the professional and technical workforce. The combination is warranted because doing so offers a better sense of the importance of technical work in the contemporary economy and avoids the problem of having to decide who is a technician and who is a professional.

**BOX 4.3**
**Occupations Counted by the Bureau of Labor Statistics
as a Professional Specialty or as a
Technician Occupation**

## PROFESSIONAL SPECIALTY OCCUPATIONS

Engineers
    Aeronautical and astronautical engineers
    Chemical engineers
    Civil engineers, including traffic engineers
    Electrical and electronics engineers
    Industrial engineers, except safety engineers
    Mechanical engineers
    Metallurgists and metallurgical, ceramic, and materials
        engineers
    Mining engineers, including mine safety engineers
    Nuclear engineers
    Petroleum engineers
    All other engineers
Architects and surveyors
    Architects, except landscape and marine
    Landscape architects
    Surveyors
Life scientists
    Agricultural and food scientists
    Biological scientists
    Foresters and conservation scientists
    Medical scientists
    All other life scientists
Computer, mathematical, and operations research occupations
    Actuaries
    Computer systems analysts, engineers, and scientists
        Computer engineers and scientists
        Computer engineers
        Database administrators, computer support
            specialists, and all other computer scientists
        Systems analysts
    Statisticians
    Mathematicians and all other mathematical scientists
    Operations research analysts
Physical scientists
    Chemists
    Geologists, geophysicists, and oceanographers

Meteorologists
Physicists and astronomers
All other physical scientists
Social scientists
    Economists
    Psychologists
    Urban and regional planners
    All other social scientists
Social, recreational, and religious workers
    Clergy
    Directors, religious activities and education
    Human services workers
    Recreation workers
    Residential counselors
    Social workers
Lawyers and judicial workers
    Judges, magistrates, and other judicial workers
    Lawyers
Teachers, librarians, and counselors
    Teachers, preschool and kindergarten
    Teachers, elementary
    Teachers, secondary school
    Teachers, special education
    College and university faculty
    Other teachers and instructors
        Farm and home management advisors
        Instructors and coaches, sports and physical training
        Adult and vocational education teachers
        Instructors, adult (non-vocational) education
        Teachers and instructors, vocational education and training
    All other teachers and instructors
    Librarians, archivists, curators, and related workers
        Curators, archivists, museum technicians, and restorers
        Librarians, professional
    Counselors
Health diagnosing occupations
    Chiropractors
    Dentists
    Optometrists
    Physicians
    Podiatrists
    Veterinarians and veterinary inspectors
Health assessment and treating occupations
    Dietitians and nutritionists

**BOX 4.3   Continued**

Pharmacists
Physician assistants
Registered nurses
Therapists
   Occupational therapists
   Physical therapists
   Recreational therapists
   Respiratory therapists
   Speech-language pathologists and audiologists
   All other therapists
Writers, artists, and entertainers
   Artists and commercial artists
   Athletes, coaches, umpires, and related workers
   Dancers and choreographers
   Designers
      Designers, except interior designers
      Interior designers
   Musicians
   Photographers and camera operators
      Camera operators, television, motion picture, video
      photographers
   Producers, directors, actors, and entertainers
   Public relations specialists and publicity writers
   Radio and TV announcers and newscasters
   Reporters and correspondents
   Writers and editors, including technical writers
All other professional workers

## TECHNICIANS AND RELATED
## SUPPORT OCCUPATIONS

Health technicians and technologists
Cardiology technologists

Clinical laboratory technologists and technicians
Dental hygienists
Electroneurodiagnostic technologists
EKG technicians
Emergency medical technicians
Licensed practical nurses
Medical records technicians
Nuclear medicine technologists
Opticians, dispensing and measuring
Pharmacy technicians
Psychiatric technicians
Radiologic technologists and technicians
Surgical technologists
Veterinary technicians and technologists
All other health professionals and paraprofessionals
Engineering and science technicians and technologists
Engineering technicians
Electrical and electronic technicians and technologists
All other engineering technicians and technologists
Drafters
Science and mathematics technicians
Technicians, except health and engineering and science
Aircraft pilots and flight engineers
Air traffic controllers and airplane dispatchers
Broadcast technicians
Computer programmers
Legal assistants and technicians, except clerical
Paralegals
Title examiners and searchers
All other legal assistants, including law clerks
Programmers, numerical, tool, and process control
Technical assistants, library
All other technicians

will return to this debate when we discuss how the lines of demarcation across these traditional categories are blurring.

## The Increasing Growth in Professional and Technical Work

Four related trends account for the expansion of the professional and technical labor force: corporate growth, the commercialization of scientific knowledge, demographic changes, and technological advances.

### Corporate Growth

Organizations, like individuals, consume professional services, and corporate growth has both directly and indirectly increased the demand for professional services. As organizations have become more numerous over the 20th century, corporate demand has augmented individual demand for professional services, thereby enlarging the market for professionals. In some occupations, such as law and accounting, corporate demand has surpassed individual demand (Derber and Schwartz, 1991). Corporations have also discovered that it is sometimes cheaper, if not more effective, to provide for themselves some of the expertise that they formerly purchased from solo practitioners or professional firms. Thus, corporations widely hire their own professionals, which has further increased demand.

Corporate growth has affected the demand for professional services indirectly as well. Aside from engineers, most professionals at the turn of the century worked as solo practitioners or in small partnerships. Doctors, lawyers, and accountants served clients from homes or offices and played economic roles similar to small business owners. Over the 20th century, solo practice dwindled. Between 1931 and 1980, self-employment among physicians fell from 80 percent to about 50 percent (Derber and Schwartz, 1991). Similarly, less than one-third of all lawyers in the United States now work as private practitioners, whereas in 1950 over half were so employed (Spangler, 1986). Even in relatively rural areas, professional services are now dispensed by law firms, accounting firms, hospitals, urgent care centers, and other professional bureaucracies that hire professionals as salaried employees.

Professional bureaucracies create employment opportunities for professionals in two ways. First, because hospitals and other professional bureaucracies have access to more resources than do solo practitioners, they can afford equipment and facilities that enable them to provide services that clients could not otherwise obtain. The availability of such services, in turn, increases the population's demand for professionals' expertise, thereby enabling professional bureaucracies to support more practitioners per capita than would have worked under a regime of solo practice. Second, professional bureaucracies create an organizational context supportive of specialization. Because professional bureaucracies colocate practitioners, they can employ specialists and still provide broad expertise to clients. However, providing breadth of expertise under a regime of specialization means that more practitioners must become involved in a case. Thus, as specialists replace generalists, more professionals are required to meet a client's needs.

### Commercialization of Scientific Knowledge

The increasing economic importance of scientific knowledge is the second reason for the expansion of the professional and technical workforce. One researcher has estimated that, by the 1960s, scientific output was doubling every 6 to 10 years, a rate of growth "much faster than that of all nonscientific and nontechnical features of our civilization" (Price, 1986:141). He also noted that 90 percent of all scientists who ever lived are alive today. The explosive growth of science has been sustained, in part, by the realization that scientific and technical knowledge can generate considerable profit. The commercialization of chemistry and physics during the late 19th and early 20th centuries gave rise to the industries on which the U.S. economy currently pivots: aerospace, automobiles, energy, pharmaceuticals, petrochemicals, and electronics. Advances in the life sciences, especially in immunology, microbiology, biophysics, and biochemistry, underwrote the expansion of the health care industry that began after World War II. More recently, molecular biology and its associated technologies have opened opportunities for entirely new industries and have revolutionized others (Teitelman, 1989; National Research Council, 1986a). The explosion of scientific activity, both basic and applied, created a wellspring of demand for scientists, engineers, technicians, and health professionals. However, the commercialization of science did not simply enlarge existing fields; it triggered a proliferation of new technical occupations via two processes: specialization and the elimination of low-level work tasks.

As the stock of knowledge in a discipline becomes more complex, scientists and related professionals find it difficult to remain generalists. Although generalists may be effective at screening problems, they are less prepared than specialists to advance a field's knowledge or provide state-of-the-art services. Because the latter activities are highly valued in technical cultures, most of the sciences and professions have adopted a strategy of carving cognate areas into ever narrower subfields. Specialization increases the number of professionals by opening up new territory and by requiring collaboration: under a regime of specialization, few individuals can execute alone tasks that require both breadth and depth of expertise.

Overburdened professionals have also sought to curb workloads by allocating routine duties to other groups. Many of the technician occupations that have flourished in the latter half of the 20th century originated in the giving off of "dirty" work by the established professions. The phenomenon has been most visible in the health care industry: licensed practical nurses and an expanding array of technicians have coalesced into occupations around tasks discarded by their more prestigious colleagues (Hughes, 1958). The dynamic is also prevalent in other industries, where it has given birth to a plethora of technical occupations ranging from the well-known—paralegals, electronics technicians, chemistry technicians—to the obscure—test and pay technicians (see Kurtz and Walker, 1975).

## Demographic Change

Increasing life spans and the upward shift in the age distribution have also contributed to the increasing prominence of professional and technical work. As people age, they require more health and social services. A significant proportion of these services are delivered by doctors, nurses, health care technicians, social workers, and other professional and technical occupations. Between 1996 and 2006, the Bureau of Labor Statistics expects professional and technical health and social service occupations to create 1.9 million new jobs. Since BLS estimates that all professional and technical occupations will generate 5.8 million jobs over the next decade, health and social service occupations will account for 33 percent of the growth (Silvestri, 1997).

## Technological Change

Perhaps the most important reason for growth in the professional and technical labor force is technological change. New technologies have shifted the workforce toward professional and technical occupations in several ways. The first has been by generating entirely new occupations. Throughout history technologies have spawned occupations: the wheelwright, the blacksmith, the machinist, the automobile mechanic, and the airline pilot are illustrations. In the past, technologies created occupations across the entire division of labor. Although modern technologies have also sired occupations in all strata, those with high technical content appear to have become more common (Adler, 1992). Commentators usually credit this change to the advent of the computer. In 1950, few people worked with computers, and most who did were mathematicians (Pettigrew, 1973). By the 1970s, computers had given birth to such well-known occupations as programmer, systems analyst, operations researcher, computer operator, and computer repair technician. These occupations, which now employ over 1.8 million workers, continue to be among the fastest growing. By 2006, computer-related occupations are anticipated to provide employment for 3 million people or 2 percent of the labor force (Silvestri, 1997).

The explosion of occupations directly related to the computer, however, is only the most visible sign that technology may now favor the professional and technical workforce. Numerous professional and especially technician occupations have been created over the last four decades by technologies other than the computer: air traffic controllers, nuclear technicians, nuclear medical technicians, broadcast engineers, technical writers, and materials scientists are examples. Although many technician occupations that have arisen de novo center on the maintenance of a technology, not all of them do. For instance, sonographers rarely repair ultrasound equipment, yet their work arose with the use of ultrasound in medical imaging (Barley, 1990). Technicians who monitor the controls of nuclear power plants and EEG technologists are further examples of occupations spawned by new technologies that have little role in the technology's maintenance.

Ironically, technological change has also augmented the professional and technical labor force by automating blue-collar and lower white-collar jobs. Computerized technologies typically automate the most routine parts of a job simply because routines are easier for designers to program. To successfully deskill or eliminate workers via automation, firms must also reallocate the more complex aspects of a target occupation's work to another occupation. Since the occupations that benefit from such reallocations tend to acquire cognitive and technical responsibilities, deskilling unintentionally expands the number of technical workers. For instance, Smith (1987) has argued that a reallocation of tasks once exercised by craftsmen and foremen was largely responsible for the birth of such technician occupations as rate-fixers, estimators, and inspection and planning engineers. Similar arguments have been made for the rise of programmers and schedulers in machine shops (Braverman, 1973).

Even when skills are not reallocated, automation may still skew a firm's labor force toward technical and other highly skilled employees, if the employment of unskilled and semiskilled labor declines disproportionately (Spenner, 1995). Several researchers have shown that two decades of computerization have altered the mix of jobs in the insurance and banking industries by precisely such a path (Baran, 1987; Attewell, 1987, 1992). Although office automation enabled firms to reduce their reliance on lower-level clerks, the relative importance of more highly skilled workers (particularly those who program and maintain computers and databases) increased as the number of clerical employees fell. Thus, when computerized automation occasions layoffs among lower-skilled workers, it leaves in its wake a work structure more heavily weighted toward members of professional and technical occupations. For similar reasons, the trend toward downsizing, even in the absence of computerized technology, should have favored the increasing importance of professional and technical work, since professionals and technicians are less likely to be downsized than are middle managers and members of other occupational categories.

## Implications of the Growth of Professional and Technical Work

The expansion and proliferation of recognized professional and technical occupations, combined with the possibility that other lines of work are becoming increasingly technical, carry tremendous implications for vertical (autonomy/control) and horizontal (task scope) divisions of labor and for the importance of cognitive and interactive skills among the labor force.

### Vertical and Horizontal Dimensions

Professional and technical workers have never fit easily into the hierarchical structures of most bureaucratic organizations (Whalley, 1986; Freidson, 1973). As Weber (1968) long ago noted, the legitimacy of a vertical division of labor is premised on the assumption that expertise can be nested: superiors ought to be able, at least in principle, to understand the work that their subordinates perform. But unless a professional's superior is also a member of the same profession, the superior's ability to understand and thereby legitimately direct or control the subordinate's work becomes suspect. Historically, three organizational forms have emerged to manage discontinuities between hierarchies of authority and expertise.

The first is the staff department. Locating professionals in staff positions removes them from the vertical division of labor and places them in the role of advisers to top executives. Organizations often use staff positions to manage lawyers and information systems specialists, for instance. The difficulty is that staff positions are only viable when the number of professionals is relatively small and when their work is not central to the organization's primary line of work.

Second, some organizations in which professional and technical work is crucial to the primary mission create two hierarchies: one to manage administrative issues and other to coordinate technical matters. Most hospitals and many research and development laboratories have such a structure. A hospital's administrative hierarchy handles such tasks as admissions, billing, procurement, and personnel, and the medical hierarchy, which is managed by physicians, coordinates patient care. Note, however, that the organizational logic of the medical hierarchy differs substantially from the organizational logic of a bureaucratic hierarchy. Whereas the nodes of a bureaucratic hierarchy are composed of positions, the nodes of a medical hierarchy are typically composed of occupations: doctors, nurses, and technicians. Occupational hierarchies not only tend to be flatter than bureaucratic hierarchies, but also they are more loosely coupled with respect to task performance. Doctors certainly set the agenda for a patient's care, but they often do not possess the specific skills or knowledge of a nurse or a technician. Accordingly, nurses and technicians are typically given considerable leeway to perform their duties as their occupational expertise dictates. As Derber (1982), Bailyn (1985), and other observers of professional work in organizational contexts have repeatedly noted, in an occupational hierarchy, superiors may be able to set goals but they are less capable of dictating practice.

The third organizational form, which Stinchcombe (1959) called "craft administration," is currently best exemplified by the construction industry. Construction projects are typically managed by a contractor who takes responsibility for coordinating the project. The contractor, in turn, usually subcontracts aspects of the project to tradespersons: plumbers, carpenters, masons, and electricians. Although the contractor sets timetables, secures resources, and ensures that the work meets specifications, the contractor does not manage the actual details of how the work is performed. The latter is the domain of the trades. Craft forms of administration are notable because they function by all but eliminating hierarchy. Although craft coordination is most well known among the trades, which are typically considered blue-collar occupations, it is also common in research and development laboratories, in advertising, in filmmaking (Faulkner, 1983) and in other forms of artistic work that have a project structure. There is evidence that project organization is spreading to other organizational contexts, in part, because of the spread of professional and technical work. The parallels between craft, professional, and technical forms of organizing highlight a crucial and often overlooked implication of the spread of professional and technical work: the increasing prominence of horizontal divisions of labor.

A horizontal division of labor implies a dispersion of authority among experts from distinct occupational groups. The logic behind this way of dividing work and authority is that knowledge and skills are domain specific and too complex to be fragmented and nested; thus individuals, rather than positions or jobs, become vessels of expertise. Knowledge is preserved and transmitted through extended training rather than through the rules and procedures that characterize bureaucracies. Occupational groups retain authority over their own work, while interacting with members of other groups to manage their respective components of a task. In a horizontal division of labor, knowledge and skills tend to be transportable across work sites. Prior to the industrial revolution, except for the military and the church, horizontal divisions of labor were the primary forms of organizing. It was only with the development of the factory system that vertical divisions of labor become more prominent.

A resurgence in the horizontal division of labor should pose problems for organizations and individuals quite distinct from those they have faced in the recent past. One such problem concerns the nature of careers. Research has long demonstrated that relatively few scientists, doctors, or lawyers desire careers structured around hierarchical advancement. The same is true for most engineers. Although the literature frequently suggests that engineers desire managerial careers and although there can be no doubt that engineers move into management at greater rates than do members of other professions (Perrucci, 1971; Ritti, 1971; Zussman, 1985; Whalley, 1991), surveys of engineers nevertheless routinely indicate that two-thirds of all engineers are more interested in careers that involve increasing technical challenge (Bailyn and Lynch, 1983; Allen and Katz, 1986). In a series of studies of technicians, Zabusky and Barley (1996) report that most technicians also aspire to "careers of achievement" rather than "careers of advancement."

The difficulty for both technicians and engineers is that organizations have historically not supported careers of achievement for three reasons. First, the range of work that most organizations can offer is not sufficiently broad to provide most technical workers with substantive challenges that are occupationally meaningful. Second, offering careers of achievement requires that organizations and human resource managers understand how members of an occupational community conceptualize skill and then plot opportunities in those terms. Such career paths are difficult to envision if one is not also a member of the occupation, which is certainly the case for most managers. Finally, definitions of success in most organizations, if not society as a whole, are still defined in terms of advancement. Thus, to pursue careers of achievement, technical professionals must often move from organization to organization in search of new challenges and skills. The desire for careers of achievement partially explains the growing popularity of contract work among technical professionals. Thus, as the professional and technical labor force expands, one might well expect independent contracting to become increasingly common.

**Cognitive Work**

Professional and technical work has always required considerable cognitive and analytical skill and familiarity with esoteric bodies of knowledge. There is little reason to believe that this will become any less or any more common among most established professional and technical occupations. Far more likely is the possibility that scientific and technical developments will require professionals to learn new bodies of knowledge continually in order to avoid becoming obsolete. Technological obsolescence has long been a concern among scientists and engineers; advanced technologies may increase the risk in other professions as well. For instance, Barley (1990) demonstrated how the advent of computerized medical imaging has inverted the status hierarchy in most radiology departments, because younger radiologists were more likely than older radiologists to understand how to interpret ultrasound, CT, and MRI images.

As work becomes increasingly technical, however, we can expect analytical skills to become more important for an ever-larger segment of the workforce and even in lines of work that have not historically been viewed as requiring analytical skills. For instance, the digitization of control systems and the integration of previously discrete subcomponents of production systems requires operators to begin to coordinate production systems via a symbolic interface instead of relying on immediate sensory data such as tastes, smells, and sights (Zuboff, 1989). Symbolic interfaces require individuals to work with increasingly abstract representations of phenomena. However, it would be a mistake to believe that technical work eliminates the need for contextual knowledge of materials and techniques, as some analysts have suggested. What is more likely is that work will increasingly require a more complex interweaving of analytical and contextual knowledge that the culture's current system for classifying work has difficulty accommodating. Technicians' work illustrates the point.

Most technicians work at an interface between the material world and a world of representations (Barley, 1996b). Using sophisticated techniques and technologies, they transform aspects of the material world into symbols that can be used for other purposes and by members of other occupations, typically professionals or managers. For instance, in medical settings, technicians produce images, counts, and other data useful for medical diagnosis. Technicians in nuclear power plants and other automated facilities create and monitor flows of information on production systems. Science technicians reduce physical phenomena to data or "inscriptions" from which scientists construct arguments, papers, and grants (Latour and Woolgar, 1979). Technicians, however, do more than generate representations and information; most are also responsible for taking care of the physical entities they oversee. Technicians ensure that machines, organisms, or other physical systems remain intact and in good working order. Care taking often requires technicians to make use of the very representations they create. Thus, emergency medical technicians take action on the basis of diagnoses made at the site of an accident. Microcomputer support technicians use the results of tests and probes to alter the functioning of computer systems.

The dual processes of transformation and caretaking that define the core of technical work also make it culturally anomalous. To be effective troubleshooters at an empirical interface, technicians must comprehend the principles of the technologies and techniques they employ as well as be familiar with more abstract, systematic bodies of knowledge. In this respect, technicians resemble professionals. For instance, emergency medical technicians require knowledge of biological systems, pharmacology, and disease processes to render diagnostically useful information. And because technicians manipulate entities to achieve practical ends, they must also possess extensive contextual knowledge of their materials, technologies, and techniques. This knowledge is usually highly situated, acquired through practice, and difficult to articulate, much less codify. Contextual knowledge resides in an acquired ability to read subtle visual, aural, and tactile cues where novices see no information at all. At present, our culture prefers to divide work neatly into mental and manual occupations. We do not have culturally meaningful categories for making sense of work that requires both. As a result, organizations typically either treat technicians as junior professionals or as a modern variant of blue-collar labor, both of which misrepresent the tech-nician's role in a production system and sometimes undermine their effectiveness (Barley, 1996b).

### Interactive-Emotion Work

The ability to manage human relationships is key to the work of many professionals. Doctors, lawyers, teachers, and social workers are expected to be particularly adept at ministering to their clients' emotional as well as their physical, legal, and educational needs. In fact, one of the primary complaints about professionals is that they often treat clients too distantly, as a mere case. Engineers and other technical professionals are even more maligned for their inability to relate to others. Increasingly, however, even the work of engineers and programmers requires considerable interpersonal skills. Technical work is today typically performed in the context of a team whose members must not only coordinate with each other but also must communicate effectively with clients in order to develop reasonable understandings of the requirements of the systems they design and build.

Thus expansion of professional and technical work is likely to necessitate that an even larger proportion of the workforce possess effective interpersonal skills. A shift to a horizontal division of labor will place an increasing premium on the ability of people with different types of expertise and information to collaborate and coordinate. Without widespread interaction and communication skills, organizations will become increasingly ineffective because managers will be unable to coordinate the details of work processes, since they will not possess the requisite substantive knowledge. In a world of technical and professional experts, systems of command and control can not ensure coordination, the communication of requisite information, or effective problem solving. In a world of experts, such activities require the involvement of specialists who know the capabilities as well as the limits of their own expertise and therefore how to work effectively in teams in order to assemble breadth of knowledge from the deeper knowledge of distributed experts. Team-based professional and technical work also demands effective emotional labor.

## Summary

Professional and technical workers have always enjoyed considerable autonomy, and we expect that this will generally remain the case. However, the employment of professionals by organizations, such as the employment of physicians by managed-care firms, may make some professionals increasingly subject to bureaucratic controls. For example, physicians in health maintenance organizations are typically required to prescribe drugs from an approved formulary, a bureaucratic limitation on their professional autonomy.

The scope of these workers' tasks is influenced by the gradual expansion of professional and technical work within organizations in which occupations collaborate. Technicians and professionals do not fit easily into the vertical blue-collar/managerial divide that dominated the rise of large-scale factory systems. In-

stead, because they hold the knowledge and expertise that is derived from their formal education and training and craft-based
experience, the boundaries of their expertise tend to be determined by the scope of their training and experience—i.e., by the
horizontal boundaries of their craft or profession. Thus, the placement of and interactions across these horizontal boundaries becomes a more salient feature for professional and technical
workers and the organizations that employ them.

Cognitive complexity is a central feature of professional and
technical work, but its substance is shifting quickly in many occupations. Draftsmen have moved from skilled constructors of material images to operators of software for computer-aided design.
Technological changes in engineering have blurred the category
between some electrical and mechanical engineers:
"mechatronics" fuses digital and mechanical systems. Social
changes have forced civil engineers to incorporate concern for and
develop expertise in environmental impact assessment.

Increasingly, the use of cross-functional teams requires professional and technical workers to have the cognitive and interactive skills needed to communicate, negotiate, and solve problems
across traditional horizontal boundaries.  In the helping professions such as nursing and medicine, workers have become more
astute at managing the emotional encounter between patients and
caregivers, understanding that healing may have as much to do
with emotional as with physical care.  Many other professionals
and technicians that work with customers and clients are now
being trained in the emotional labor aspects of their work.

## CONCLUSIONS

Given the heterogeneity subsumed by any aggregate of occupations, attempting to draw conclusions about the general directions in which the content of work may be changing within a
broad occupational classification is a risky proposition.  It is even
riskier to make proclamations about the direction in which the
workforce as a whole is moving.  Most of the data on the changing content of work comes from case studies of specific organizations, ethnographies of particular occupational groups, and a
handful of surveys that assess a wider population of workers,

albeit much more superficially than do situated studies. For some lines of work, especially management, there is almost no research on which to draw. In these instances, students of the changing nature of work must, at present, turn to accounts written by consultants and managers and then attempt the difficult interpretive task of separating self-serving rhetoric from substantiated observations.

Ideally, researchers and policy makers should be able to turn to a continually updated, national, and longitudinal database on job skills and occupational structures to identify with less ambiguity the kinds of trends with which this committee has wrestled. Unfortunately, it is precisely the absence of such a database that occasioned this report in the first place. We do not believe, however, that the absence of systematic and consistent data on a large number of occupations precludes drawing any conclusions whatsoever. Although the data are uneven, accumulating evidence concerning some occupational groups does seem to point in consistent directions. It is reasonable to treat these consistencies as a form of replication that constitutes reasonable evidence. And it is equally important to underscore where there is considerable variation in what researchers have observed as well as to indicate what we do not, but should, know.

At the moment, evidence for what is happening to blue-collar jobs, especially jobs in manufacturing settings, is the most well developed and consistent. Our relatively greater understanding of the changing content of blue-collar work probably reflects several facts. First, social sciences have a long history of studying factory work, so there are baselines for judging change. Second, it is much easier for researchers to gain access to blue-collar settings than it is to gain access to any other type of work, with the possible exception of clerical, technical, and professional work. Third, techniques for describing and analyzing physical work are better developed than techniques for describing and studying mental and interpersonal work. Finally, the transformation of production processes has been a key concern in business and engineering since the early 1980s and hence has attracted considerable attention from responsible journalists.

Aside from the well known fact that blue-collar employment has fallen precipitously since mid-century, four other develop-

ments seem reasonably widespread. Although these developments are occurring unevenly across industries and although one
can surely find within any particular industry instances to the
contrary, compared with the past an increasing number of blue-
collar jobs seem (a) to offer workers more autonomy and control
over their work processes, (b) cover a wide range of tasks, (c)
demand more interpersonal skill, and (d) possibly have become
more analytic if not cognitively complex. The adoption of lean
production techniques, the growing acceptance of team-based
work systems, and the spread of computer-integrated manufacturing technologies appear to be primarily responsible for these
changes in the content of blue-collar jobs. Of these various developments, the least well-documented concerns the implications of
computer-integrated manufacturing systems for analytic skills. In
particular, we don't know whether the increasing importance of
symbolically mediated work is confined to particular types of jobs
or whether it is associated with specific types of production systems. We do know that the best evidence for the increasing analytic complexity of blue-collar work comes from studies of
computerized control systems in continuous process industries:
steel, chemicals, and paper manufacturing.

Developments in the content of service work are far less consistent. For instance, studies of service work sometimes indicate
a reduction in autonomy and control, less cognitively complex
tasks, a narrowing of task scope and more routinized and scripted
interpersonal interactions. Other studies indicate precisely the
reverse. Our sense is that this variance is not an artifact of the
studies that have been published, but an accurate reflection of
what is happening in the service industries. Although social scientists have long studied clerical work, few other service occupations have attracted the attention of researchers until recent years.
As a result, our ability to differentiate between types of service
work is poorly developed. If nothing else, recent research indicates the utility of developing more grounded concepts for conceptualizing different types of service work. Researchers have
shown that the ability to make even rudimentary distinctions,
such as a typology of customers or the difference between relational and transactional interactions, greatly improves the ability
to identify trends, at least within subsets of service work.

Nevertheless, despite situational variance in the findings of existing studies, several tentative conclusions regarding service work seem plausible. First, a significant percentage of service jobs are probably becoming more routinized, in large measure because new information technologies enable greater centralization and control over work activities. Second, there is a tendency toward the blurring of clerical and sales jobs. Although the heterogeneity of work within specific service occupations appears to be increasing, this heterogeneity reflects, at least in part, the tendency to structure work differently according to market segments. Finally, interpersonal and emotional skills remain critical to service work, although the nature of these skills varies and the language for describing them is primitive.

The nature of most professional and technical work is probably not changing in dramatic ways, even though scientific and technical advances can completely alter what professional and technical workers need to know in a relatively short span of time. Professional and technical jobs continue to afford considerable autonomy and control over work processes, to have considerable scope, to offer cognitively complex challenges, and to demand high levels of interpersonal skills. The primary trends in professional and technical work are continued expansion and increasing specialization, which tends to generate even more professional and technical occupations. When taken together, professionals and technicians now represent the largest occupational sector of the U.S. labor force. This constitutes a significant change in the demography of employment in the United States and other Western economies.

Finally, solid empirical evidence on changes in the content of managerial work is nearly nonexistent. There is no broad occupational sector that demands more careful empirical study. This situation is ironic given the amount of attention paid to managerial work in the business press and by consultants and educators in business schools. The dearth of research on the changing nature of managerial work probably exists for several reasons. First, many descriptors used to inventory managerial skills (for example, makes decisions, engages in planning, etc.) are insufficiently detailed and too imprecise to register evidence of change. Second, the language for making distinctions among managerial

jobs is poorly developed, perhaps, in part, because an intense belief in the importance of general management skills has precluded
attempts to map functional specialization in management and to
inventory the kinds of skills that such specialization requires. The
situation is analogous to claiming that there are no meaningful
differences between the work of a neurosurgeon and the work of
dermatologist because both are doctors. Finally, as most ethnog-
raphers of work know, it is notoriously more difficult to gain access to observe the work of managers than it is to gain access to
study the work of those whom managers supervise.

Nevertheless, the tremendous consistency in the business
press's portrayal of how management is changing indicates that
substantial changes may indeed be occurring in the nature of
managerial work. These changes appear to have been occasioned
by the same developments that have altered blue-collar work:
namely, downsizing and the shift to team-based work systems.
Two developments seem especially plausible, although they are
in need of much better documentation and they are likely to ex-
hibit considerable variation across firms, industries, and hierarchical levels. First, at least lower-level managers appear to have
experienced some loss in authority and control. Second, the need
to communicate horizontally across the internal and external
boundaries of organizations may be becoming more important
than the supervision of an employee's work. There is also considerable talk about the substantive content of managers' jobs, shifting toward the procurement and coordination of resources,
toward coaching as opposed to commanding employees, and toward project management skills. In attempting to assess these
changes, however, it is particularly difficult to separate rhetoric
from reality.

Given the variation of developments within and between
broad occupational groups and the paucity of research on service
and especially managerial work, it is difficult to draw many
strong conclusions regarding general trends in the nature of work.
Nevertheless, considering all available evidence, the committee
believes two conjectures concerning the broader trajectory of
work in a postindustrial economy seem particularly plausible and
worthy of considerably more scrutiny. First, it does not appear
that work is becoming more routine or less skilled than in the

past, but we are unwilling, at present, to claim that the reverse is true. Second, and far more intriguing, is the increasing importance of what sociologists call the horizontal division of labor. By a horizontal division of labor, sociologists tend to mean an occupational division of labor in which expertise is distributed among groups of specialists. In a vertical division of labor, expertise is lodged in organizations and structured in series of proper subsets that form an inclusion tree or a hierarchy in which superiors know what subordinates know and more. The tendency toward the reduction of job categories and the increasing scope of work in blue-collar work, the expansion and proliferation of professional and technical jobs, the segmentation of service work by problem area and market, and the hypothesis of increasing specialization of managerial work all point to a more horizontal system for organizing tasks, skills, knowledge, and responsibility. In horizontal divisions of labor, coordination occurs though the ongoing collaboration of experts rather than through a system of command and control. Should the general nature of work change to favor a more horizontal division of labor, it would represent a reversal of one of the primary attributes of the industrial era: the primacy of bureaucracy and hierarchy. In an economy marked by a horizontal division of labor, content and knowledge would become more important than command and control as vectors for organizing.

# 5

# Implications for Occupational Analysis Systems

The changes in work that we have described in this book pose two major challenges to occupational analysis systems. First, the external and organizational contexts and the content of work are changing. Second, the full scope and direction of the changes are not well known because, in part, we lack the data needed to track and assess the consequences of the changes that are occurring. What is needed is an occupational analysis system that tracks changes in the nature of work in a way that assists in both projecting future conditions and designing new jobs. This is what we mean by forward-looking, using historical data to both project the future and influence design decisions.

In this chapter, we therefore explore two questions: How can occupational analysis systems support efforts to both track and assess the changes in work occurring now and in the future? And how can occupational analysis systems support organizational and individual planning, counseling, and decision-making processes to adapt to these changes and achieve the outcomes from work that are critical to them?

More specifically, we ask: What are the implications of the changing world of work for occupational analysis tools and methods, and for occupational structures? How can occupational analysis and occupational structures best remain current, relevant, and useful? Can the changes be successfully addressed by existing systems? If not, how must occupational analysis and classification systems be designed to better address the changes? What are the implications (if any) of the failure of occupational analysis to adapt to changes at work for the performance of the institutions that use occupational analysis?

*Occupational analysis* refers to the tools and methods used to describe and label work, positions, jobs, and occupations. Among the products of occupational analysis is an *occupational category system*, or an *occupational structure.* Our use of these terms is closest to that of organizational and industrial psychology, and it is somewhat different from the use of terms by other communities and disciplines. For example, sociologists and economists would ascribe additional meaning to the term *occupational structure*, including patterns of occupational recruitment and retention and inter- and intragenerational patterns of occupational mobility. For us, *occupational classification* has two general meanings: (a) the act of classifying positions, jobs, or occupations into an existing occupational category system and (b) the set of occupational categories in an occupational category system.

Occupational structures reflect the nature of work, its organization, employment relationships, demographics, and other factors. They also reflect their intended purposes and influence, both directly and indirectly, the variety of outcomes depicted in Figure 1.1. Occupational structures are the lenses through which we categorize and view the system of work. Over time, they also help *shape* the system of work by providing the labels and categories that we use to bundle tasks and duties into positions, jobs, and occupations—in effect telling analysts, employers, and recruiters what is salient about work and what is not.

For example, organizations that rely heavily on existing occupational classification systems and categories for the recruitment of personnel may be less likely to identify new task mixtures in their existing job structure, and also less likely to import new occupational distinctions from other organizations in the same industry. Similarly, organizations that use category systems that afford little or no attention to teamwork features of work organization may lag in the adoption of such structures and the monitoring of their effectiveness. Occupational structures serve a defining function that tends to be backward-looking, reflecting what existed in the past, rather than forward-looking, reflecting trends in the changing organization of work.

Choices of methodology and technology for occupational analysis are guided by theories of work and occupations, represented most clearly by existing occupational structures. The primary consideration in occupational analysis is to devise a system that provides a basis for understanding the world of work, one that is grounded in this reality. Any arbitrary structure will not suffice. The relevance of a system issues from its connection to reality (recognizing that the structure and the reality can never be fully separated), along with the extent to which it serves its intended purposes. To the extent that reality is changing or undergoing significant shifts, an occupational analysis system should be able to measure the change, and the categories should reflect it.

We begin our discussion with a brief history of occupational analysis systems; then we describe contributions of two different types of systems—descriptive and enumerative—and assess how well the most current system under development performs the two key functions of occupational analysis: tracking changes in work and supporting employment decisions and career counseling. In our analysis we examine the extent to which existing and prototype systems of occupational analysis systematically address the major themes of heterogeneity detailed throughout this volume: increasing heterogeneity of the work-force, increasingly fluid boundaries between who performs which jobs, and the increasing range of choices around how work is organized and structured.

## HISTORY

The development and evolution of occupational analysis systems has been closely tied to wars and other major social changes. Most observers note that occupational analysis systems evolved principally in response to one or another practical personnel problem, and they have often involved a key role for government in their initiation and definition (see Primoff and Fine, 1988; Mitchell and Driskill, 1996). For example, following the Civil War, the early attempts at civil service reform were aimed at more orderly placement of people in federal jobs to overcome the political spoils system (Primoff and Fine, 1988). At about the same time, the U.S. Census Bureau began to do more formal grouping and analysis of occupational titles, beyond mere listing (National Research Council, 1980). In their historical account, Mitchell and Driskill (1996) noted the widespread misuse of personnel during World War I, due primarily to a lack of definition of job requirements. The response of the U.S. Army was to commission leading psychologists to improve personnel testing and placement. Among the results were the Army Alpha and Beta tests for selecting and classifying recruits, as well as other occupational analysis efforts after the war aimed at improving the match between people and jobs.

In the period between the two world wars, there also were major developments in occupational analysis and category systems. The U.S. Civil Service Commission launched major new efforts in the early 1920s to analyze a comprehensive set of jobs and occupations in terms of their duties, requirements, and advancement prospects (Mitchell and Driskill, 1996). The Wagner-Peyser Act of 1933, passed by Congress at the depths of the Great Depression, established the U.S. Employment Service with the basic aim of helping workers find suitable jobs. The act also established an extensive occupational research program. This research endeavor, closely coordinated with the Social Science Research Council and the National Research Council, eventually produced the first edition of the *Dictionary of Occupational Titles*, with subsequent editions produced by the Employment Service in 1949, 1965, 1977, and 1991. During this period, key research communities in industrial psychology were formed, including such figures as Sidney Fine, Ernest McCormick, Ernest Primoff, and Carroll Shartle, individuals who would later develop occupational analysis methodologies that substantially inform the current state of the art.

Occupational analysis methodologies underwent further developments in and around World War II in both the military and the civilian sectors, as well as in their intersection. For example, during World War II the War Manpower Commission could sanction firms for labor pirating (Jacoby, 1985:262). That commission also asked firms in the civilian sector to classify jobs into the categories defined in the recently released *Dictionary of Occupational Titles* (DOT). The goals included providing better training and more systematic ways to transfer labor to needed areas. Social and behavioral scientists who did occupational analysis played key roles in the development of more sophisticated personnel selection, training, and promotion systems in occupations ranging from Air Force pilots to Navy submarine crews. Similarly in the civilian sector, by 1947 an estimated 20 percent of U.S. industrial firms used employment tests for hiring and placement (Jacoby, 1985).

The development of occupational analysis after World War II is interwoven with: (a) the dramatic expansion of higher education, including the number of sociologists and economists specializing in the analysis of occupations and labor markets, and research and practice communities in industrial and organizational psychology, vocational guidance, and employment training; (b) the expansion of survey research activities facilitated by improved sampling and item analysis procedures as well as by the advent of statistical data processing via computers; and (c) the expansion of employers' efforts to systematize hiring, training, promotion, and compensation systems.

Most of the major systems described in this chapter had their origins in this period. That includes the DOT, the Standard Occupational Classification System (SOC), the Occupational and Employment Statistics Classification System (OES), and the Military Occupational Specialties (MOS) system as it is underpinned by the Comprehensive Occupational Data Analysis Program (CODAP), which was not fully implemented until the 1960s (Mitchell and Driskill, 1996). Beginning in the late 1950s, a number of occupational analysis systems were developed for use in the private sector, many of which built on work completed in the military and government sectors (Fleishman, 1967, 1992; McCormick et al., 1972; McCormick, 1979; Cunningham et al., 1971; Cunningham, 1988).

The SOC has recently undergone revision. The new system will be used by all federal agencies to collect occupational data; it will provide the occupational classification system for the 2000 census; and it will be used for coding jobs in the latest revision of the DOT, known as O*NET™. Although there is no external crisis, such as war or depression, driving the development of O*NET™, this new system offers a response to many of the weaknesses in previous systems. At present, O*NET™ is at the prototype stage of development; moving it to an operational system will require a strong sponsor. In the committee's judgment, O*NET™ holds significant promise for dealing with changing work contexts and content. In a later section of this chapter, we provide a detailed description of its key features. Appendix A presents additional details and a discussion of its prototype development and evaluation.

## TYPES OF OCCUPATIONAL ANALYSIS SYSTEMS

A useful way to organize a review of occupational analysis systems is to distinguish between systems that emphasize occupational categories and subsequent enumeration, and systems that emphasize descriptive analysis of the content of work. The characteristics of these two types of systems are quite different; the differences arise because their purposes, historical roots, and methods of development have differed. Wootton succinctly states this (1993:3-9):

> As was the case for the U.S. [*Dictionary of Occupational Titles*], other national dictionaries [descriptive analysis systems] were developed from the "bottom up" through expensive and extensive job analyses, mostly with enterprises. Specification of detailed occupations was weighted towards manufacturing and production occupations that were dominant in the early post-war period. Some countries, particularly English speaking nations, tended to borrow other countries' dictionaries before they had their own. Initial versions of dictionaries usually were developed between the 1940's and the early 1970's.
>
> Most statistical occupational classification systems [enumerative] were developed to serve the needs of population censuses. These structures usually were developed from the "top down" according to analytical principles. Occupational categories tend to be fewer in number and broader than those in dictionaries, and they provide little information other than occupational title, alternate titles, and a brief description of tasks. Occupational coding is based on these items. Aggregation principles often appear to be heterogeneous

within the same system—a mix of tasks performed, function, industry, and education or training required.

Figure 5.1 presents an abstract model that shows a useful way to conceptualize the relationship between descriptive analytic systems and category/enumerative systems. The rows in this matrix represent occupational categories. These categories can and do vary in their specificity, representing quite specific jobs in some systems (e.g., police detective) and relatively broad occupational families in others (e.g., public safety occupations). The columns in the matrix represent requirements or other characteristics that are descriptive of the categories or rows, for example, the knowledge and skills required by occupations (geography, biology, negotiation, troubleshooting), characteristics of the environment of occupations (work schedule, indoors/outdoors, presence of hazardous materials), and work activities completed in occupations (analyzing data or information, handling and moving objects, assisting and caring for others). The cell entries (e.g., cell 1,1; cell 3,2) are numerical values or other information that denotes the standing of each occupation on each attribute. For example, if attribute 1 is "presence of hazardous materials" and category 1 is "police detective," then the value in cell 1,1 might be a rating of the frequency with which police detectives encounter hazardous materials in the course of doing their jobs.

| Occupational Categories (e.g., occupations, occupational families) | | Occupational Attributes (e.g., required work activities, skills, knowledge, number of incumbents, compensation) | | | | |
|---|---|---|---|---|---|---|
| | | Att. 1 | Att. 2 | Att. 3 | . . . . . | Att. ZZZ |
| | Cat. 1 | Cell1,1 | Cell1,2 | Cell 1,3 | | Cell 1,ZZZ |
| | Cat. 2 | Cell 2,1 | Cell 2,2 | Cell 2,3 | | Cell 2,ZZZ |
| | Cat. 3 | Cell 3,1 | Cell 3,2 | Cell 3,3 | | Cell 3,ZZZ |
| | . . . | | | | | |
| | Cat. N | Cell N,1 | Cell N,2 | Cell N,3 | | Cell N,ZZZ |

FIGURE 5.1 Matrix depicting conceptual relationship between categories of occupations and attributes of occupations.

By conceptualizing the category/enumerative and descriptive analytic systems as the rows and columns, respectively, of a matrix, a more complete and integrated view of occupational analysis and structure is possible. Furthermore, if a hierarchy is imposed on both the rows and columns of such a matrix, allowing the aggregation and assembly at differing levels of specificity, then it is possible to think of a complete occupational analysis/structure system that can meet varying user needs.

Our discussion is divided into three sections. The first section presents an overview of the descriptive/analytic systems; the second presents category and enumerative systems; and the third presents systems that combine the enumerative and descriptive approaches (e.g., DOT, O*NET™, MOS). Appendix B presents additional details of most of the systems mentioned.

## Descriptive Analytic Systems

As noted above, descriptive analytic systems have as their primary purpose the detailed description of occupations in terms of a number of attributes. These systems often do not concern themselves at all, or only secondarily, with a particular structure or category system of occupations. The systems discussed below are primarily descriptive, borrowing their categorical structures from elsewhere or creating such structures for particular applied purposes, sometimes employing the descriptive data obtained through their system to create the structure. Table 5.1 contains summary information about six illustrative descriptive analytic systems. These systems, taken together, represent the long stream of research on occupational analysis arising out of the Great Depression and World War II eras, as well as more recent developments that attempt to capitalize on the lessons learned through that research. They illustrate the utility of descriptive analytic systems that are appropriate for a large proportion, if not all, occupations through the use of a set of common attributes for describing occupations. These systems rely on a variety of data sources described in Appendix B. They are listed below:

- *Position Analysis Questionnaire*: This is perhaps the best-known example of a worker-oriented job analysis technique

TABLE 5.1 Type of Jobs, Descriptors, and Applications for Some Illustrative Descriptive Analytic Systems

| Name | Types of Jobs | Types of Descriptors | Applications (lists are illustrative, not exhaustive) |
|---|---|---|---|
| Position Analysis Questionnaire (PAQ) | All | Information input; mental processes; work output; relationships with others; job context; job demands | Selection of employees; job evaluation, grouping and design; performance appraisal; position classification; job matching |
| Fleishman Job Analysis Survey (F-JAS) | All | Abilities—cognitive, physical, psychomotor, sensory/perceptual and social-interactive; job skills and knowledge | Job descriptions; selection of employees; classification (people into jobs); performance appraisal |
| General Work Inventory (GWI) | All | Activities—sensory, information-based, physical, interpersonal; general mental and physical requirements; work conditions; job benefits | Job description and grouping; selection and placement of employees |
| Common Metric Questionnaire (CMQ) | All | Background; contacts with people decision-making; physical and mechanical activities; work setting | Job description and evaluation; performance appraisal; position classification |
| Multipurpose Occupational Analysis Systems Inventory-Closed Ended (MOSAIC) | All federal jobs | Tasks; competencies; personal and organizational styles | Position description; position classification; selection of employees |
| Work Profiling System (WPS) | Managerial or professional; service or administrative; manual or technical | Job tasks; job context | Selection and placement of employees; performance appraisal; job design, description, and classification |

NOTE. This table is adapted with permission from Peterson and Jeanneret, 1997:36–44.

(McCormick et al., 1969). It has a long history of development and research and its strengths and weaknesses are well known. The questionnaire has 187 items listing work behaviors and job elements at a level of abstraction that permits work to be described across a broad range of occupations.

• *Fleishman Job Analysis System*:   This system is likewise based on a long history of research, primarily in the area of human ability testing (Fleishman and Quaintance, 1984). A unique contribution of this system was the development of behaviorally anchored rating scales to assist subject-matter experts (job incumbents, supervisors, or other persons knowledgeable about the jobs to be rated) in estimating the amount of each ability required to perform a job or job task (Fleishman, 1992).

• *General Work Inventory*:   This inventory grew out of the research on the occupational analysis inventory (Cunningham et al., 1990; Cunningham, 1988) and had its origins primarily in occupational education and guidance.

• *Common Metric Questionnaire*:   This is a more recently developed "worker-oriented" job analysis instrument intended to apply to a broad range of jobs and to overcome some of the perceived inadequacies of earlier systems, particularly the relatively difficult reading level of descriptor items and the relative (as opposed to absolute) nature of the ratings obtained for jobs (Harvey, 1991).

• *Multipurpose Occupational Systems Analysis Inventory-Closed Ended (MOSAIC)*:   This system, recently developed by the Office of Personnel Management, is designed to collect and distribute data about tasks and competencies (combinations of knowledge, skills, and abilities) for occupations within large occupational families (Gregory and Park, 1992).

• *Work Profiling System*:   This system likewise uses instruments tailored to occupational families rather than a single instrument intended to apply across all jobs in the workforce (Saville and Holdsworth Ltd. USA, Inc., 1990).

## Category/Enumerative Systems

This section describes international systems, national systems outside the United States, and major U.S. systems. In the discussion of U.S. systems, we focus attention on the current revision plans for the Standard Occupational Classification. This revision promises a number of important advances over existing systems, including the provision for assimilating new occupations into the system on an ongoing basis.

## International Standard Classification of Occupations

The International Standard Classification of Occupations (ISCO) has three objectives: (1) to facilitate international communication about occupations through its use internationally; (2) to provide international occupational data for research, decision making, and other activities; and (3) to serve as a model, but not a replacement, for countries developing or revising their national occupation classifications. The development of this structure was based on the recommendations and decisions of the Thirteenth and Fourteenth International Conferences of Labour Statisticians, held at the International Labour Office, Geneva, in 1982 and 1987. The underlying source data consist of population censuses, statistical surveys, and administrative records maintained at the national level. ISCO-88 is the 1988 revision of the 1968 version of the classification system (International Labour Office, 1990).

The ISCO system uses two key concepts: job and skill. Job is defined as "a set of tasks and duties executed, or meant to be executed, by one person." Skill is defined as "the ability to carry out the tasks and duties of a given job." The ISCO-88 structure is hierarchical, with 10 major groups at the top, 28 submajor groups, 116 minor groups, and 390 unit groups.

Although ISCO-88 was not intended to be the single structure that would fit all nations, many nations have adopted the ISCO system with little or no modification (Elias, 1993). Western nations have tended to make more substantial modifications to ISCO-88 or to devise their own structures (Wootton, 1993). Australia, Canada, the United Kingdom, and the Netherlands are primary examples. Table 5.2 presents comparative information about ISCO-88 and its adaptation for use in those countries. Shown are the key classification concepts, the number of levels in the hierarchical ladder, and the number of occupational groups in each level. All of the systems have fewer than a dozen broad

TABLE 5.2  Key Concepts and Occupational Groups per Hierarchical Level for Some National Occupational Classification Systems

| System | Key Concepts | Number of Occupational Groups Per Level in Hierarchy (broadest level to most discrete level) | | | |
|---|---|---|---|---|---|
| ISCO-88 | Job<br>Skill level<br>Skill specialty | 0<br><br>1 | 28 | 116 | 390 |
| Australia | Job<br>Skill level<br>Skill specialty | 8 | 52 | 282 | 1,079 |
| Netherlands | Job<br>Skill level<br>Skill specialty<br>Main task<br>Specific skills | 5 | 43 | 121 | 1,211 |
| United Kingdom | Job<br>Skill level<br>Skill specialty | 9 | 22 | 77 | 371 |
| Canada | Job<br>Skill level<br>Skill type<br><br>Major work performed | 10 (NOC)<br>10 (SOC) | 26 (NOC)<br>47 (SOC) | 139 (NOC)<br>139 (SOC) | 522 (NOC)<br>514 (SOC) |

NOTE: Canada employs two slightly different classification structures; see appendix for details.

groups at the top of the hierarchy, and 26 to 52 groups at the next level. In all cases but one, nations have developed systems with more units at the lowest level than are found in ISCO-88. Some of these nations have tried to be closely compatible to ISCO-88 and others have revised their existing systems, by using classification principles modeled on ISCO-88 or reasonably similar to those principles.

Elias (1993) and Wootton (1993) drew some lessons for the United States based on their experiences with and review of these and other classification systems. Elias cautioned against spending too much money on a new system, urging revision rather than complete rebuilding. His other points of counsel included: establish a long-term program rather than attempt a one-shot revision, use criteria for revising the classification system but be prepared to ignore them in the interests of practicality, use a committee of producers of occupational statistics and description to make the revisions but have consultation with users, be prepared to handle the advocacy of specific occupational groups, seek advice from industry-based organizations, expect strong academic criticism of any revision effort, put resources into automated coding and user-friendly guides, and, finally, know that both the enumerative and the descriptive systems are essential and both must be maintained. In agreement with Elias, Wootton does not think ISCO-88 is useful for U.S. needs. First, it offers no added value over existing U.S categorical systems; second, its use of skill level and type as an aggregating principle has produced a grouping structure that is inconsistent across nations. However, she believes it "provides some useful general guidelines" for the U.S. revision process—one is the move toward "replacement of multiple, fragmented classification systems with a unified structure, another is the shift in the basis of classification towards an explicit recognition of occupational skills" (1993:329).

## Major U.S. Systems

Several enumerative systems are used in the United States:

- the Classified Index of Occupations and Industries (U.S. Department of Commerce, 1990a),
- the Occupational Employment Statistics Structure (U.S. Department of Labor, 1992),
- the Standard Occupational Classification (U.S. Department of Commerce, 1980).

In the past, a patchwork of cross-walks has been required to link the occupational categories across these systems (National Occupational Information Coordinating Committee, 1993), and the quality and currency of these cross-walks are uneven. The latest revision of the SOC is expected to alleviate the need for these cross-walks.

**U.S. Bureau of the Census Classified Index of Occupations and Industries**  Employment information is collected as part of the decennial census. Questions are asked of a sample of the census population about the job title of each family member and the duties of the job. Their responses are coded by Census Bureau staff using its Classified Index of Occupations and Industries. Since the census data are available only every 10 years, the Current Population Survey (CPS) was established; this survey is conducted monthly by trained interviewers who visit a carefully selected sample of households to obtain responses from a household member, whose responses are coded by Census Bureau staff using the alphabetical Index of Industries and Occupations (U.S. Department of Commerce, 1990b). Dempsey (1993) characterizes the CPS data as, until recently, the largest and single most important source of comprehensive data on employment by occupation and still the primary source of local data and information on worker characteristics. He lists the strengths of the CPS as historical continuity (over 50 years of monthly surveys), provision of detailed information about adult workers, and the use of interviewers, which provides the opportunity to explain the meaning of questions and otherwise enhance the quality of the data collected. However, he notes that the CPS sample is too small to provide data on specific occupations on a monthly basis.

*Occupational Employment Statistics (OES)*  The OES program is conducted by the Bureau of Labor Statistics and state employment security agencies to obtain counts of wage and salary employment in all nonagricultural establishments. The Bureau of Labor Statistics provides technical guidance and support as well as survey design, and the state agencies collect the data from business establishments. The survey is conducted so that the entire economy is covered within a 3-year cycle. The classification system used by the OES is compatible with the SOC, but not identical.

*Standard Occupational Classification (SOC)* The SOC was first released in 1977 and was intended to provide a mechanism for cross-referencing and aggregating occupation-related data collected by social and economic statistical reporting programs. This work began in 1966 to address a long-standing need to establish a *single* occupational classification system for governmental agencies. As noted above, the SOC has been revised and is intended to be the primary occupational category system used by all federal agencies (Office of Management and Budget, 1997).

All federal agencies that collect occupational data will use the new system; similarly, all state and local government agencies are strongly encouraged to use this national system to promote a common language for categorizing occupations. The new SOC system will be used by the OES program of the Bureau of Labor Statistics for gathering occupational information. It will also replace the Census Bureau's 1990 occupational classification system and will be used for the 2000 census. In addition, the new SOC will serve as the framework for information being gathered through the Department of Labor's Occupational Information Network (O*NET™), which is in the process of replacing the *Dictionary of Occupational Titles*.

The revision process of the SOC began with the establishment of the Standard Occupational Classification Revision Policy Committee (SOCRPC). The SOCRPC used the Bureau of Labor Statistics' OES system, together with O*NET™, as the starting points for the new SOC framework. It laid out 10 criteria for revising the SOC (Box 5.1). Although these are similar to the principles outlined for the original system, there are some important differences. The scope of the SOC is extended to include "all occupations in which work is performed for pay or profit"; only occupations unique to volunteers are excluded. Wording was added to the second principle, reflecting that the structure should be flexible enough to assimilate new occupations as they become known, a feature of particular importance in light of the changing context and content of work. A new principle, the third, indicates the need for a linkage with past systems. The fourth principle significantly extends the kinds of factors that would be used to classify occupations, from just work performed, to work performed plus skills, education, training, licensing, and credentials.

---

**BOX 5.1**
**Standard Occupational Classification:**
**Criteria for Revised System**

1. The Classification should cover all occupations in which work is performed for pay or profit, including work performed in family-operated enterprises by family members who are not directly compensated. It should exclude occupations unique to volunteers.

2. The Classification should reflect the current occupational structure of the United States and have sufficient flexibility to assimilate new occupations into the structure as they become known.

3. While striving to reflect the current occupational structure, the Classification should maintain linkage with past systems. The importance of historical comparability should be weighed against the desire for incorporating substantive changes to occupations occurring in the work force.

4. Occupations should be classified based upon work performed, skills, education, training, licensing, and credentials.

5. Occupations should be classified in homogeneous groups that are defined so that the content of each group is clear.

6. Each occupation should be assigned to only one group at the lowest level of the Classification.

7. The employment size of an occupational group should not be the major reason for including or excluding it from separate identification.

8. Supervisors should be identified separately from the workers they supervise wherever possible in keeping with the real structure of the world of work. An exception should be made for professional and technical occupations where supervisors or lead workers should be classified in the appropriate group with the workers they supervise.

9. Apprentices and trainees should be classified with the occupations for which they are being trained, while helpers and aides should be classified separately since they are not in training for the occupation they are helping.

10. Comparability with the International Standard Classification of Occupations (ISCO-88) should be considered in the structure, but should not be an overriding factor.

SOURCE: *SOC Federal Register Notice*, 1998:3.

---

The eighth principle acknowledges the decrease in the hierarchical nature of work by noting an exception to the principle of separately identifying supervisors from those they supervise. For professional and technical occupations, supervisors or lead workers are to be classified *with* the workers they supervise. The suggested changes should enhance the capacity to effectively capture changes in work.

With publication of the revised SOC on July 7, 1997, the SOCRPC invited public comment and suggestions for revision. More than 200 responses were received. As a result, the SOCRPC published another revision on August 5, 1998, with a significantly modified hierarchical structure and numbering system. In the proposed revision of the SOC, there are four major levels of aggregation: (1) major group (designated by the first two digits of the SOC code), (2) minor group (designated by the third digit of the SOC code), (3) broad occupation (designated by the fourth and fifth digit of the SOC code), and (4) detailed occupation (the sixth digit of the SOC code). The proposed structure contains 810 detailed occupations, 449 broad occupations, 98 minor groups, and 23 major groups. At the minor group level, older versions of the SOC are compatible with the Census 3-digit occupations, but not identical.

This structure is more differentiated than the 1980 SOC, which enumerated 665 occupations. According to the 1997 version of the revision, the number of occupations in computers, design, science, health, law, education, and arts increased from 192 to 286 (about a 50 percent increase), whereas the number of mechanical and production occupations decreased from 246 to 157 (a decrease of 35 percent). These changes are in agreement with the trends in the changing content of work noted in Chapter 4. Military occupations are listed with their civilian counterparts in this revision, except for 20 military occupations for which no civilian counterpart could be found. These occupations make up the military occupations group.

The SOC Revision Policy Committee, noting that it has been 18 years between revisions of the SOC, recommended the establishment of a standing committee, the Standard Occupational Classification Review Committee, to maintain the currency and appropriateness of the SOC to the world of work. We regard the adoption of a uniform occupational coding system as a critical development with distinct policy implications. On balance, a uniform system is more desirable than the current collection of discrete systems strung together with cross-walks made at various times by various agencies. However, because the needs of potential users vary, some flexibility in its use should be maintained. Furthermore, we concur with the recommendation of the SOC Revision Policy Committee to form a standing body to perform such functions as continuous updating. This would relieve the need for cross-walks for rapidly changing or emerging occupations. Likewise, deficiencies in the system for particular consumers of occupational information could be routinely identified and handled, again mitigating the need to form new structures for consumer needs not properly addressed in the decennial revisions.

### Systems Combining Descriptive and Enumerative Features

### Dictionary of Occupational Titles

The *Dictionary of Occupational Titles* (DOT) was developed by the U.S. Employment Service to provide a catalogue of the occupational titles used in the U.S. economy as well as reliable descriptions of the type of work performed in each occupation. The latest revision of the fourth edition of the DOT was issued by the Department of Labor in 1991 (U.S. Department of Labor, 1991 [1972]). In this revision, the key distinction was a concentration on occupations in industries that had undergone "the most significant change" since 1977. In almost every other respect, there was no change: the basic concepts (e.g., element, task, position, job, and occupation) were defined as before; each of over 12,000 occupations was still "defined" by seven items: occupational code, title, industry designation, alternate titles, body of the definition (lead statement, task element statements, "may" items), undefined related titles, and definition trailer.

The DOT occupational code has nine digits. The first three digits indicate a particular occupational group, the second three digits provide the "data, people, and things" codes for the occupation, and the final three digits differentiate an occupation from all other occupations. When there are two or more occupations in a group (as defined by the first three digits of the code), then they are assigned codes alphabetically, in multiples of four. Occupations are grouped by "similarity" according to the information contained in the body of the definition, i.e., the lead statement and the task element statements.

The data in the DOT are collected by occupational analysts in U.S. Employment Services offices using methods described in the *Handbook for Analyzing Jobs* (U.S. Department of Labor, 1991). Essentially, the occupational analyst observes and interviews workers on the job and then completes a job analysis report. The handbook contains instructions for completing this report, including lists of codes and anchors for rating scales to rate data, people, and things. These dimensions are the crux of functional job analysis, a methodology developed by Sidney A. Fine during the 1950s (Fine and Wiley, 1971); it begins with the premise that all worker-job situations are an interplay of the worker with the dimensions of data, people, and things. The purpose of job analysis is to ascertain the *functionality* of the relationship of the worker to data, people, and things—that is, the complexity of interplay with elements needed to accomplish task objectives. In addition, the fourth edition of the DOT has associated data on just over 60 variables. These include training time, aptitudes, interests, temperaments, and various physical demands and working conditions. Most of the variables are binary in form. The last time the variables were systematically updated was in the middle 1970s, and some would argue (National Research Council, 1980) that this update was minimal, hence the data refer to the world of work in the middle 1960s. This is a serious limitation in a world of work that is changing.

Evaluations of the DOT lauded its value and high level of use by government agencies, researchers, and others concerned with occupational information, but they tended to criticize its unwieldy size and the growing disparity between its definitions and the real world of work (National Research Council, 1980; Spenner et al., 1980; Advisory Panel for the *Dictionary of Occupational Titles*, 1993). The DOT's principal weaknesses were identified:

• Jobs are described at a level of job-specific detail that makes it difficult to conduct cross-occupation comparisons. Thus, it would be difficult to determine to what extent the incumbents of one occupation would have the necessary general skills for easy transition to other work. As a result, the goal of using the system as a resource for workers transitioning from obsolete or down-sized occupations could not be realized.

• Jobs are described solely according to tasks. Information about the skills, abilities, knowledge, and other individual qualities needed to perform jobs is not directly collected. The latter information may be crucial to answer questions inherent in person-job matching, training, skill transfer, and wage and salary administration.

• The DOT provides some information about the physical and ergonomic aspects of jobs, such as noise, temperature, and work schedules. However, other contextual factors, such as interpersonal demands and stressors, organizational influences, and exposure to other hazards, are not covered.

• The time and expense involved in updating descriptive job information ensure that a substantial portion of the information in the DOT is outdated at any given time.

• The discrete, qualitative descriptions in the DOT do not allow for linkages with other occupational or labor-market databases.

This evaluation led to the development of a prototype to address these weaknesses and replace the DOT. This replacement was named the Occupational Information Network or O*NET™.

## Occupational Information Network (O*NET™)[1]

The Occupational Information Network— O*NET™ — is an electronic database of information, rather than a book. In it, the information about each occupation has been considerably expanded and the number of occupations included in the system

---

[1]The material in this section was heavily borrowed from the Occupational Information Network (O*NET) Technical Executive Summary (American Institutes for Research, 1997).

has been considerably reduced. We believe that O*NET™, when fully developed and given adequate support and maintenance, will provide a useful tool for tracking changes in work, assisting in job design, and in supporting employment decisions. The following discussion of O*NET™ is divided into three sections—a description of the components of the content model, a general statement regarding prototype evaluation, and a brief presentation of the electronic database and sample screens. More detail on the evaluation is found in Appendix A.

*The Content Model* The system was conceptualized as a way to aid a wide range of users, such as job applicants, career counselors, training specialists, displaced workers, vocational rehabilitation counselors, recruiters, state and federal labor and manpower specialists, and public- and private-sector employers. The system was designed to address an impressive array of tasks, including:

- determining aptitude and skill requirements for jobs,
- assessing a person's suitability for an occupation,
- developing training standards and competency standards for jobs,
- comparing the skills required for a displaced worker's previous and prospective jobs, and
- documenting physical and contextual demands of jobs.

Although it may seem difficult to envision an occupational information system capable of serving so many goals and the needs of so many users, this was nonetheless the design goal for O*NET™.

The content model developed for O*NET™ is based on three key postulates (Peterson et al., 1995, 1999). First, jobs can be described quantitatively according to variables that generalize across jobs. For example, they may be described in terms of inductive reasoning or the physical requirements that apply to many jobs. The content model is designed to be a general, reasonably stable descriptive system. Second, multiple windows (organizing systems) can be used to observe the world of work. Each window reflects a set of descriptors associated with an applied use of the system. For example, skills and knowledge are of primary interest when one is concerned with specifying training needs, whereas selection and placement decisions are more likely to be made on the basis of abilities and educational credentials. Multiple windows allow the system to address multiple applications. Third, within a given domain of descriptors, variables can be organized hierarchically. Hierarchical arrangement of descriptors allows users to access multiple levels of specificity and provides a way to organize job-specific descriptors, such as tasks, within a more general cross-job structure. The content model covers six domains:

- worker characteristics,
- worker requirements,
- experience requirements,
- occupational requirements,
- occupation-specific requirements, and
- occupational characteristics.

Broadly speaking, the content model assumes that jobs can be described according to either the demands placed on the people doing the work (worker-oriented descriptors) or the work being done (occupational descriptors). The first three domains relate to worker-oriented requirements. Workers bring to the job certain characteristics, such as abilities and interests, and, as a function of their experiences, develop certain capacities that help them do the job. Worker requirements also include the skills and knowledge people must acquire to do the work. The last three domains relate to the work people do—occupational requirements. These are described by generalized work activities, for example operating heavy equipment, and by job-specific tasks, such as the steps involved in operating a specific type of forklift. These work activities, however, are influenced by requirements imposed by the job environment or work context, as well as requirements imposed by the organizational structure or context. As such, these contextual variables are also subsumed in the occupational requirements domains. Finally, all of these variables, in turn, interact with the global features of the organization and its operating environment; such occupation characteristics make up the final

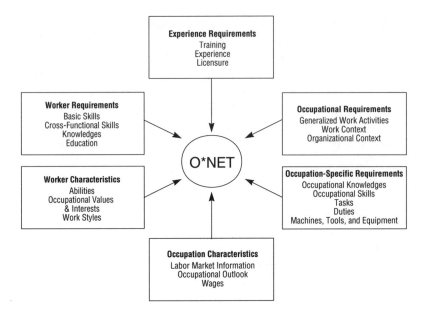

FIGURE 5.2 Descriptive domains of the O*NET™ content model.


domain. The six domains contained in O*NET™'s content model are depicted in Figure 5.2 and summarized below.

*Worker Characteristics* Worker characteristics reflect relatively enduring features of the individual that might influence job performance. They include: (1) abilities, (2) values and interests, and (3) work styles. The *ability* constructs are predominantly drawn from Fleishman's ability requirements taxonomy and include basic cognitive, psychomotor, physical, and perceptual abilities, virtually all of which are known to have direct relevance to many jobs. *Interests* are described under the rubric of Holland's six-factor taxonomy, a typology that is extensively used in the career counseling literature and in conjunction with the current *Dictionary of Occupational Titles*. The six orientations include realistic, investigative, social, artistic, enterprising, and conventional (Holland, 1985). *Occupational values* are described using the 21 descriptors of the Occupational Values Questionnaire, which is based on the Minnesota Job Description Questionnaire (MJDQ)

(Dawis and Lofquist, 1984). The MJDQ is a unique values-type measure, in that it describes occupations relative to the needs they reinforce. These include such noncognitive, stylistic reinforcers as authority, creativity, security, and variety. Finally, *work styles* refers to personality characteristics that are either directly relevant to job performance or that may facilitate development of requisite knowledge and skills. The taxonomy of work styles was drawn from recent efforts to formulate models of adaptive personality characteristics encompassing five to nine factors—the seven first-order constructs in the model are achievement orientation, social influence, interpersonal orientation, adjustment, conscientiousness, independence, and practical intelligence.

**Worker Requirements** Worker requirements are the *skills* and *knowledge* that people develop as a function of education, practice, and experience. Skills are broadly defined as sets of general procedures that underlie the effective acquisition and application of knowledge in various domains of behavior. Thus, skills are viewed as more tractable and less stable than abilities, in that they can be enhanced more quickly through practice, and as more proximal to effective work performance than abilities, in that they are usually more closely associated with particular types or classes of tasks. The general skills and knowledge in the O*NET™ system are thought to be transferable across jobs and thus should play a progressively more important role as organizations seek to develop a workforce capable of adapting to new types of job demands.

*Skills* are organized into six broad categories likely to be involved in virtually all jobs. The first is *basic* skills—developed cognitive capacities that allow for learning or knowledge acquisition. Basic skills are divided into *content skills*, such as reading, listening, oral and written communication, and declarative knowledge, such as mathematical procedures, and *process skills*, such as critical thinking, learning strategies, and application of principles. The remaining five categories are termed *cross-functional skills*, skills that facilitate performance across a variety of settings. They include problem solving, social skills, technological skills, system skills, and resource management skills. In addition to basic and cross-functional skills, the occupation-specific skills of each occupation must also be considered. They facilitate work across a variety of settings.

Knowledge is defined as a collection of discrete but related and original facts, information, and principles about a certain domain. Accordingly, types of knowledge were identified by determining basic types of concepts likely to be applied in a variety of different jobs (i.e., basic concepts involved in electronics, psychology, and transportation were considered, among other areas). In all, 33 broad knowledge areas were identified, with each area subsuming a number of more specific concepts. For example, biology might subsume cellular biology, ecology, genetics, and biochemistry.

*Experience Requirements*    Experience requirements refer to training and career history events that influence knowledge and skill development. In contrast to skills, abilities, and knowledge, which describe the actual capabilities and competencies that an individual may bring to a job, variables in this domain either provide evidence that the person has actually performed the same or similar work previously (experience), or provide evidence that the person does or did possess the knowledge and/or skills necessary to perform the job (education, training, or licensure). Relevant work experience refers to job tenure in related jobs and training experiences in the work context, including apprenticeships, on-site training, and on-the-job training, whereas educational and licensure requirements focus on the amount and type of education or licensure required. The information in the experience domain may be used differently by various O\*NET™ users. Those seeking jobs or career guidance would most likely be interested in the amount and type of education or licensure required, whereas potential employers would be most interested in evidence of competence signified by completion of a particular educational or training regimen.

*Occupational Requirements*    Three major areas are included within the occupational requirements domain—generalized work activities, work context, and organizational context.

*Generalized work activities* are defined as an aggregation of similar job activities/behaviors that underlies the accomplishment of major work functions. These broad types of job activities occur to different degrees in a very broad spectrum of occupations. The 42 generalized work activities included in O*NET™ were identified through earlier factor analyses of job analysis inventories, with generalized work activities subsumed under four general categories:   information input, mental processes, work output, and interacting with others inside and outside the organization.

*Work context* variables describe the conditions under which job activities must be carried out. They include physical conditions (e.g., temperature and noise) as well as social psychological conditions (e.g., time pressure and dependence on others) that might influence how people go about performing certain activities. Although some occupations are carried out across a wide range of settings, many others can be said to have a typical work context, such as indoor/outdoor work, degree of danger and exposure to elements and hazardous materials, and degree of involvement or conflict with other persons.

*Organizational context* refers to variables that might interact with the operational environment and how people go about doing their work. For example, a flatter, more open organizational structure may require workers who possess a broader range of skills, placing a premium on problem-solving skills and an independent work style. O*NET™ organizational context variables were identified after a review of studies assessing the impact of organizational structure on how work gets done, with a special focus on high-performance organizations. Other examples of organizational context variables include industry characteristics; a range of characteristics of organizational structure and human resource systems and practices; organizational values; individual and organizational goals, processes and characteristics; and various features of role relationships, including conflict, negotiability, and overload. Given that incumbents of occupations reside in a wide variety of organizational and industry contexts, the eventual O*NET™ system will need to attend to multiple sources of data and will need to measure not only typical contexts but also the variety of contexts in which occupations reside.

*Occupation-Specific Requirements* Occupation-specific descriptors such as tasks, duties, machines, tools, and so on cannot generally be derived from literature reviews; they require input from job holders and their supervisors. In the prototype O*NET™ system, tasks were the sole occupation-specific variables included, and these tasks were primarily extracted from existing DOT descriptions, supplemented by other sources. Additional kinds of occupation-specific descriptors are being generated for occupations and organized according to the broader cross-job structure. So far, research suggests that identification of job-specific descriptors can be facilitated by the availability of a broader, cross-job organizing structure. Furthermore, it appears that by organizing the specific descriptors in terms of a broader common language, it becomes possible to apply job-specific information more efficiently.

*Occupational Characteristics* Occupation characteristics refer to economic conditions that shape the nature of the organization, its market, and employment conditions. Measures of these constructs are drawn into O*NET™ by linking with databases, such as those maintained by the Bureau of Labor Statistics, for example, the Occupational and Employment Statistics Classification System and the Current Population Survey. Many but not all of these variables, such as compensation and employment projections, will be of interest to both those considering or counseling about entry into specific jobs and those doing manpower and succession planning for organizations or industry trade associations.

*Field Test and Prototype Evaluation* The current O*NET™ is a prototype for an eventual fully developed system, the goal being a comprehensive, flexible occupational information and analysis system that is national in scope and that tracks changes in a way that provides a basis for future projections and the design of new jobs. The developers of the system, along with those working on its validation now, have several years of experience with O*NET™, including initial technical evaluations and valuable lessons learned about the potential problems, prospects, and policy issues associated with a full-scale system.

An initial set of studies was conducted with the O*NET™

instruments (Peterson et al., 1996, 1999). Data were collected from job incumbents in approximately 30 of an initial targeted sample of 70 occupations. In addition, occupational analysts rated 1,122 occupations on a subset of O\*NET™ descriptors using information from the DOT to guide their ratings. Taken as a whole, the analyses of the prototype provide good initial support for the appropriateness of the O\*NET™ content model and provide some guidance on its possible utilization. However, there is still much that could be done to further this understanding. The availability of the database of analyst ratings of 1,122 occupational units made many of the analyses possible. The available incumbent datasets, however, are still relatively small and may not be representative of all 1,122 occupational units. The conclusions we draw based on the data available to date must be reevaluated when a more complete dataset of job incumbents is available. A detailed discussion of the prototype evaluation is provided in Appendix A. This includes a discussion of cross-domain analysis, grouping operations, aggregation of descriptor variables, and linking job analysis to assessment using the job component validation model.

*The Electronic Database*  The O\*NET™ electronic database was created using the data collected during the field test of the O\*NET™ questionnaires and the data provided by the analysts' ratings (Rose et al., 1996, 1999). A user interface was developed to allow users to access the data and become familiar with the system. Some of the functions that can be performed with this prototype system are illustrated in Figures 5.3 through 5.7. These illustrations apply only to the prototype software interface; it is intended that users develop specific interfaces for their particular applications. Also, the Department of Labor will issue updated databases and interfacing software as they are developed.

Figure 5.3 shows the opening screen for the O\*NET™ prototype system, a typical Windows™ interface. Buttons along the top show the various functions that can be performed with the data, including browsing, matching, filtering, exporting, and printing. The titles of the first 20 occupations in the "Incumbent" database are displayed in the "O\*NET™ Occupation Title" window; scrolling down or using the alphabetical tabs allows viewing of the rest of the occupational titles. Or one could directly

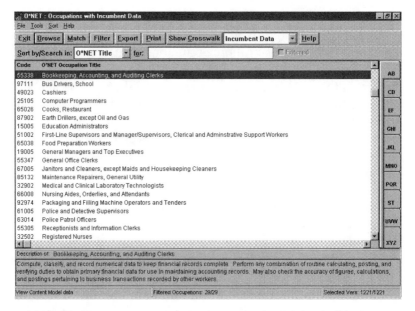

FIGURE 5.3   Opening computer display screen for the O*NET™ prototype system.

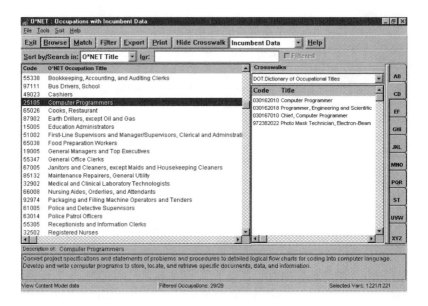

FIGURE 5.4  Computer display screen showing cross-walks across O*NET™ and DOT systems for the O*NET™ prototype system.

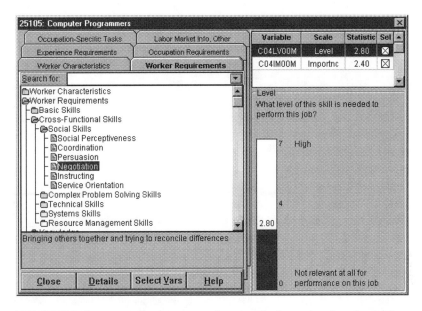

FIGURE 5.5 Computer display screen showing the browsing function of the O*NET™ prototype system.

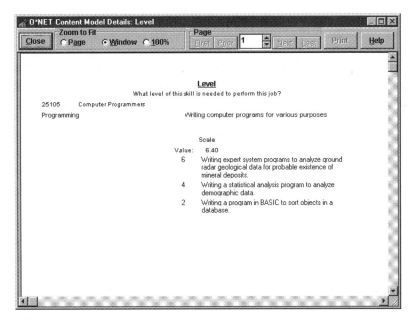

FIGURE 5.6 Computer display screen showing details of rating scale anchors for the O*NET™ prototype system.

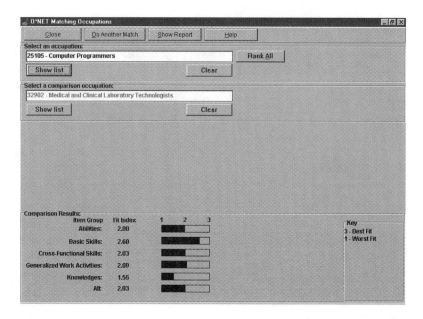

FIGURE 5.7  Computer display screen showing the match function of the O*NET™ prototype system.

search by entering the title or code of an occupation directly in the "Sort by/Search In" window. Note that the definition of the highlighted occupation is displayed in the window at the very bottom.

In Figure 5.4, the "Computer Programmers" occupational unit is displayed and the DOT codes corresponding to this unit are shown in the "Crosswalks" window, directly to the right of the O*NET™ title window. Cross-walks are available to a number of other systems.

Figure 5.5 shows the "browsing" function of O*NET™; here the "Negotiation" skill level rating for computer programmers is shown (it is 2.8 on the 7-point scale). This information was obtained by "drilling down" from the top of the O*NET™ job descriptor hierarchy, through "Worker Requirements," to "Cross-functional Skill," to "Negotiation." Note that an importance rating is also available for this skill.

Figure 5.6 shows the details of the rating scale anchors for "Programming." These are the actual anchors used by job incumbents in making the ratings of their jobs during O*NET™ data collection. A screen like this is available for any of the O*NET™ variables using level, importance, or frequency ratings.

Finally, Figure 5.7 shows the results of using the "Match" function to estimate the degree of similarity between "Computer Programmers" and "Medical and Clinical Laboratory Technologists." Similarities are computed using the level ratings, weighted by importance, for the abilities, skills, generalized work activities, and knowledge. The screen prototype shows the degree of similarity on a 3-point scale (1 = little similarity, 3 = high similarity). These pairwise comparisons are computed in fractions of a second and can be computed against all other occupations (but with more time required). This function was included in the prototype purely as an illustration of the potential capability of the O*NET™ approach; more research would be required to identify the best matching algorithm for particular purposes.

These figures depicting O*NET™ 's prototype computer display screens illustrate its potential versatility due to its sophisticated electronic medium in contrast to a static dictionary like the DOT (although the contents of the DOT have been available in electronic format for some time). The data can be explored and played with; with the development of end-user applications, a large number of user needs could be addressed.

*Advances*  In the committee's judgment, O*NET™ offers several important advances over previous systems in its job description variables and associated data collection instruments, in its electronic databases with job incumbent and occupational analyst ratings, and in the initial technical evaluations. First, O*NET™ offers the prospect of a system that brings together the most current category and enumerative systems and the most comprehensive descriptive analytical systems and makes the database readily accessible in electronic format. Coupled with its goal of being national in scope and its coverage of the economy, when fully developed it will be the first available system with these features.

Second, O*NET™ has a theoretically informed and initially validated content model with a more detailed set of job descriptors than other available systems, particularly if considered in a national context.

Third, the O*NET™ database can be accessed and used through multiple windows or modes, including not only entering with job titles or occupations at varying levels of hierarchical detail, but also entering at the level of work descriptors (i.e., knowledge, skills, abilities, other contextual factors). The latter window of access is extremely important in a world of work that is changing. It allows the user to build up inductively to the level of job or occupation, in contrast to systems that proceed deductively, starting with a job or occupational category that is nested in the past and may not be current in its ratings or job descriptive information. When fully developed, this will be a valuable feature for users wishing to depict presently emerging or newly designed jobs, or even to design a new occupation.

Fourth, O*NET™ offers a significant improvement over earlier systems, particularly DOT-based systems, in the ease of conducting cross-occupational analyses and comparisons.

Finally, the O*NET™ system, by utilizing the cross-walks supplied by the National Occupational Information Coordinating Committee, allows mapping to other major category and enumerative systems, including military classification systems. The O*NET™ system will include the revised SOC classification categories as its primary designation of lowest-level occupational units, although some of the SOC units may need to be further split—just as some of the current OES units were split in the prototype development effort. (O*NET™ occupational units were developed by analysts who estimated the similarity of occupations using a combination of DOT and OES information. The SOCRPC considered O*NET™ occupational units in revising the SOC categories, which in turn were adopted in some instances for use in O*NET™.) Given the apparent new widespread commitment of federal statistical agencies to use the revised SOC system, all such applications and data-gathering efforts would in effect be seamlessly interleaved (or nearly so) with the O*NET™ categories and database. This would allow, for example, information collected in the Current Population Survey or the Occupational and Employment Statistics Surveys to be merged, providing continuous updates of selected occupational information in the O*NET™ database.

## Military Occupational Specialties

Each branch of the military maintains its own occupational classification system, but cross-walks have been completed that relate each military occupational category to civilian counterparts in the DOT. In general, the military systems include occupations grouped within occupational families, with comprehensive lists of job tasks carried out in each occupation. Recent defense cutbacks have led to serious reexamination of the purposes and methods used by the military to conduct occupational analysis and classification (Bennet et al., 1996; Sellman, 1995). One promising alternative is to apply the methodology of O*NET™ to classifying military occupations. Chapter 6 takes up the discussion of the Military Occupational Specialty System and the application of O*NET™.

## ASSESSMENT

O*NET™, in conjunction with SOC, represents the latest developments in enumerative and descriptive systems. In this section, we focus on how well such systems capture and help track changes in work contexts and content. In the committee's view, a system such as O*NET™, when fully developed, can be effectively used to meet the needs of the changing workplace.

The changes in the world of work raise two types of issues for occupational analysis: (1) what information is gathered about work and (2) how and from whom the information is gathered. The first is referred to as content, the latter as process. Figure 5.8 depicts the changing elements of organizations and work identified in this book and guides the discussion here. This figure uses the framework provided in the introduction to summarize key findings related to environmental forces, organizational structure, worker attributes, work content and structure, and work outcomes.

There are several important factors to consider in assessing an occupational analysis system. The first is how well it incorporates the influences of environmental forces on work context and content. For example, as shown in the figure, employers have a wide range of choices in modifying their organizations to take

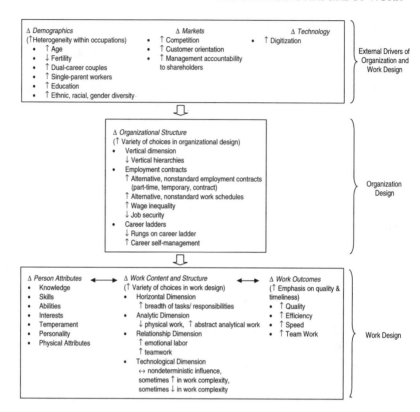

FIGURE 5.8  Changing elements of organization and work design.

into account the heterogeneity resulting from shifts in demographics, markets, and technology. An effective occupational analysis system must also be able to adequately reflect the changing content of work. As organizations shift toward self-managing teams to organize individual workers, there is a decrease in the vertical dimension and an increase in the horizontal dimension of work. As the nature of work changes, so do worker requirements—that is, the character of the job influences the attributes of the people who will fill the jobs (represented as the persons component in the figure). It is important to note that this is a two-way relationship, as the available talent pool can also influence the design of the job.

## Content Issues and Requirements

We have made the case throughout this book that work is changing: it is becoming more technical and analytical due to the increased availability of digital information technology; it is becoming more team-based due to organizational restructuring; and it is becoming increasingly customer oriented. We have also concluded that the workforce is becoming more diverse, that there are more nonstandard employment contracts, and that people are moving between occupations with more frequency than in the past. All of these changes have potential implications for the tools and methodologies of occupational analysis. Since we expect changes in work to continue, there is a need for a system that can incorporate changes as they occur.

Current changes indicate that job incumbents must have greater abstract analytical knowledge. As work is increasingly performed in teams, teamwork skills are becoming more critical. The rise of the service economy requires workers who are skilled in the domain of emotional labor. At the same time, physical labor is declining as a component of many occupations.

O*NET™ provides the most comprehensive set of person, task, and organizational descriptors of any repository of occupational analysis information (see Appendix A). The O*NET™ descriptor dictionaries were assembled after exhaustive reviews of the literature on the definition and measurement of worker and task characteristics. In the domain of cognitive abilities alone, there are seven categories (i.e., verbal, idea generation and reasoning, quantitative memory, perceptual, spatial, and attentiveness) and several specific abilities within each category (e.g., within the idea generation category, problem sensitivity, deductive reasoning, and inductive reasoning are three of seven abilities). Each person attribute is defined and accompanied by rating scales for the level and importance of the attribute for job performance. With regard to the increasing orientation toward customers and the resulting increase in jobs with emotional labor and teamwork, for example, O*NET™ offers a promising avenue for both classification and research by including an interpersonal orientation category with elements such as cooperation, concern for others, and social orientation. In the committee's judgment, the

O*NET™ dictionaries, because of their comprehensiveness, provide the best single source of descriptors for both workers and jobs.

Although the O*NET™ database is far from complete, and it remains for future research to determine if its descriptors are sufficient to capture the nuances of changes in work, the committee regards the potential of O*NET™ as superior to any alternative enumerative or descriptive occupational classification system. Its database and viewers will permit searching, sorting, and classifying on the basis of job titles, and also on the basis of hundreds of other building-block work descriptors in the content model. Although these capabilities are not exclusive to O*NET™, they are far more advanced in O*NET™ than in other occupational information systems. They are at the heart of its design and represent the real enabling potential of the system with respect to changes in the nature of work. When fully developed, O*NET™ will not only assist in new job design but will also serve as an analytic framework for monitoring change in the nature of work. The scientific state of the art of these relatively young methodologies needs to be advanced—not only do the statistical issues associated with clustering need illumination, but also theoretical and conceptual guidance is needed to inform clustering of occupations, based on these kinds of work descriptors. As evidence accumulates about the changing nature of work, difficult decisions about the implications of these changes for O*NET™ work descriptors and structure will have to be addressed.

## Process Issues and Requirements

Changes in work have implications for such issues as whom to sample when analyzing a job or occupation, how to reach them, which questions to ask, what survey technology to use, and how often to update the analysis. The increasing diversity of employment contracts, for example, raises questions about who should be considered to be a job incumbent and therefore included in the analysis; employees with all types of employment contracts who work either on- or off-site should presumably be included. As boundaries between jobs become more permeable and duties are increasingly shared among team members who have different job titles, the question of whom to ask for what information becomes more salient, as does whether the proper unit of occupational analysis is the team member or the team. Traditional approaches to job analysis commonly rely on supervisors as well as job incumbents as subject-matter experts. In many organizations today, supervisor subject-matter experts have been replaced by self-managing teams.

These issues pose challenges to occupational analysis, although most if not all of them can be met with adaptations of methodology. An important question underlying these process issues, however, is the status of job titles as indicators of work performed. O*NET™ was specifically designed to permit access to occupational information by avenues other than job titles, in recognition of the limitations of titles in today's workplace. The committee regards this as an important conceptual advance in occupational database technology and one that is important in characterizing the changing nature of work.

Other changes in work discussed in the preceding chapters have potential implications for occupational analysis processes. For example, the discussion of emotional labor in Chapter 4 also raises what may be more subtle questions about sampling procedures in occupational analysis. It is standard practice to gather task and worker attribute ratings from supervisors as well as from job incumbents. As suggested by the research literature, if emotional work processes are performed differently by men and women, or are perceived differently by male and female supervisors (e.g., emotional behaviors regarded by male supervisors as voluntary are seen by female supervisors as prescribed), then these differences are likely to appear in job analysis ratings. This in turn has potential implications for such procedural details as sampling of subject-matter experts, rating instructions, and survey contents.

## CONCLUSIONS

The quality and usefulness of particular occupational structures and data can be summarized by considering two related questions: Is the world of work adequately represented in the occupational information system in question? Can system users successfully apply the information to achieve their purposes? In this final section, we consider how these questions can be answered in terms of O*NET™, the most recent attempt to develop a comprehensive occupational information system.

## Representativeness of Occupational Structures and Data

Occupational information systems are created from data about what people do at work. The quality of an occupational information system is therefore determined by how well it represents what workers do and how their work is organized. Data about what people do at work are typically gathered using occupational analysis tools and methodologies. The quality of an occupational information system built from such data therefore rests fundamentally on the adequacy of the instruments and techniques used to gather the data and the adequacy of the samples on which the data are gathered.

Occupational information systems are derived from samples of subject-matter experts defined in various ways. O*NET™ developers gathered data from incumbents of 30 occupations employed by a stratified random sample of 1,240 establishments listed in the Dun and Bradstreet files. Whether or not this is an adequate sample depends on the inferences the developers wished to draw from the data. Before a representative sample can be drawn, the population of interest must be defined and a sampling plan for gathering data must be specified. In the absence of these, the adequacy of a sample cannot be determined and valid inferences about a population cannot be reached.

The term *validity* refers to the correctness of inferences about a population derived from observations (*measurements*) of a sample of that population. For inferences to be valid, several components must be in place: the research question must be formulated (e.g., what do working U.S. citizens do for a living?); the population must be defined (e.g., the U.S. working population); a sampling plan must be established (e.g., a random sample of 10,000 U.S. households with all working household members included); measurements must be taken that are relevant to the topic (e.g., using an established job analysis instrument); the sample must be realized (e.g., all potential respondents participate); and the measurement data must be correctly analyzed and interpreted.

Data collection to complete O\*NET™'s database will continue for the next several years. Any evaluation of the representativeness of the current database is therefore premature. If data collection for the O\*NET™ prototype was a test case of procedures for the full database, however, several implications for future data gathering can be inferred. The establishment-based sampling plan yielded only a 27 percent participation rate for employers who initially agreed to participate. In addition to remedies for increasing this rate (see Appendix A), we also suggest that the question of representativeness of the data be considered. If O\*NET™ is intended to encompass all forms of work activity in the U.S. economy, then data collection based solely on an establishment-based sampling plan will probably be inadequate. An employer-based sampling plan is likely to yield a sample of employees, that is, people holding traditional full-time jobs. It is less likely to include part-time, contract, temporary, job-sharing, telecommuting, and self-employed workers than a sampling plan based, for example, on U.S. households. To draw correct inferences about the U.S. working population from an establishment-based sample, one would have to know how the employer population differs from the working population as a whole.

Most job analyses are conducted locally, in a specific organization to support particular human resource applications. Some job analysis databases are designed to be national in scope to provide normative comparisons for client organizations (e.g., the Position Analysis Questionnaire). Historically, the U.S. Department of Labor has systematically sampled employers and jobs in order to maintain the currency of the DOT. There is thus considerable precedent for gathering occupational analysis information by sampling establishments; indeed, this has been the norm. Nevertheless, one must be careful in drawing inferences from such a sample. Occupational information contained therein probably generalizes to work as it is done in establishments that participate in such surveys (which may be most of them); it probably does not generalize to the U.S. workforce as a whole. Establishment-based samples *can* contribute information about how work is organized within organizations, a topic of increasing importance given the increasing reliance on teams and alternative employment contracts. However, the committee is unaware of any systematic establishment-based occupational surveys that gather detailed information about work organization.

If work is changing (due to different uses of technology, a greater diversity of employment contracts, demographic changes of workers, etc.), then adequately representing change in occupational information systems requires an adequate sampling plan for measuring change. At a minimum, this would require repeated measurements of a representative sample of the U.S. workforce. A cross-sectional plan repeated periodically would be adequate to identify changing trends and patterns of work. A longitudinal sampling plan including the same respondents over time would be needed to track the changing career paths of workers.

Since the nature of the sample determines the type of data collected and its usefulness for various applications, and an adequate sampling plan is critical to the next stage of O*NET™ development, in the committee's judgment, high priority should be given to evaluating the appropriateness of alternative sampling strategies.

## Usefulness of Occupational Structures and Data

Issues of the usefulness of an occupational information system like O*NET™ include the quality level of the data, enabling potential for coping with change, cross-walks among related systems, system maintenance and control, and further technological developments.

### Validity and Reliability Studies

Initial validity and reliability studies were undertaken during the first three years of O*NET™ prototype development (Peterson et al., 1996, 1999). Such research is important to establish the measurement integrity of the instruments and the reliability of the data-gathering procedures. In the committee's view, these studies should continue and should be expanded to include laboratory and field studies comparing various conditions of data collection, such as differing demographic makeup of job experts, expanded contents of questionnaires (e.g., new or additional knowledge, skill, and ability), mode of data collection (paper and pencil questionnaires versus computer-administered or Internet-administered), and studies designed to further illustrate the validity and utility of O*NET™ data for various applications. Practical issues concerning reading level and length of O*NET™ questionnaires should also be further investigated from the standpoint of gathering incumbent data for the nearly 1,100 occupations remaining.

### Scientific Quality of the Enabling Potential of the System

Because of its relational database structure, O*NET™ not only can assist in new job design/redesign but also can serve as an analytic framework for monitoring changes in the nature of work. Looking to the long-term future, one can envision a decentralized data collection design in which a multitude of individual users from across the economy—perhaps each with different purposes and level of use of the system—might contribute incumbent data on job descriptors organized into a system for identifying new jobs and work arrangements and regularly updating the information in the database. Dramatic changes in the uses of information technology in the workplace and at home increasingly bring such a system closer to the present. As with the 1998 SOC, accommodating new occupational information that has implications for content and structural changes to an existing database is a technical challenge that should be addressed by a standing O*NET™ Revision Policy Committee. We advise that such a mechanism be established.

### Coordination with Other Analysis and Classification Systems

At present there appears reasonable prospect of the coordinated use of a single occupational classification system (the 1998 revised SOC) by various federal agencies, one that would be seamlessly interleaved with O*NET™. The committee fully endorses such coordination. For mappings or cross-walks to other systems (i.e., historical, international, private-sector), it is important to know the scientific quality of the cross-walks and the locus of responsibility for their generation, maintenance, and quality control.

## Military and Civilian Systems

The military and civilian occupational analysis and classification systems are presently integrated through the use of cross-walks. The revised SOC more fully integrates military occupations into the civilian sector. However, the descriptive information on military occupations is largely restricted to lengthy lists of discrete tasks that do not generalize across occupations. Some work has been completed in the Air Force using more generalized job descriptors like those in O*NET™ to analyze military occupations (Ballentine et al., 1992; Cunningham et al., 1996). O*NET™, if used by the military services, could serve as a common language to describe military and civilian occupations, enhancing the ability to move personnel smoothly across both sectors in cases of national emergency, or even for the relatively more common task of placing veterans into civilian occupations. Some questions surrounding this issue include: How well does O*NET™ meet military needs going into the next century? What additions (or deletions) to O*NET™ would be required to make it a practical and valuable tool for the military? What synergies can be attained by using the same system in military and civilian sectors?

## Ownership, Control, and Liability

Who will own and maintain quality control over future editions of O*NET™? A federal entity? A private but federally funded entity? A private entity? What are the legal implications— if occupational classification technologies are misused, who will be liable? What are the costs of fully developing and maintaining O*NET™? Who will pay? What are the ownership, privacy, and liability implications for users and for applications that are based on O*NET™ databases and methodology? Answering these questions will be part of the full implementation of the system. We urge the U.S. Department of Labor to address and resolve these issues soon.

## Moving from Prototype to an Operational System

As we have emphasized elsewhere in this chapter, moving O\*NET™ from a prototype to a fully operational system is a significant undertaking. Data collection is paramount in this regard. Some of the more important issues in data collection include identifying samples of establishments and data sources (incumbents, analysts, or others); securing cooperation from establishments and sources; accounting for important contextual variations within occupations; and appropriately aggregating data for use in the O\*NET™ database.

These issues and data collection, in general, must be approached in a systematic manner. Clearly, there is a requirement for a long-range plan that identifies which occupations will be targeted for data collection at what times, both for collecting the original O\*NET™ data and for updating that data on a regular basis. Such a schedule can be difficult to adhere to, but it seems essential to do so if the database is to maintain integrity and credibility with users. Probability sampling of establishments and data sources is the most desirable method for ensuring the representativeness of the obtained data, and it is perhaps the best way to ensure that contextual variations are accounted for in the data. However, simple probability sampling approaches may not always be feasible. For example, the distribution of members of some newer occupations across establishments may not be well known. Also, as we note elsewhere, some occupations may not be easily accessed through establishments because individuals in those occupations are primarily self-employed. More targeted approaches may be necessary in such cases. Such approaches could include the use of unions, professional associations, business groups, or other institutions to assist in identifying samples and data sources and in securing cooperation from members of the occupation. Given that random sampling, stratified random sampling, and nonrandom targeted sampling approaches are likely to be used and that, at any given point in time, data sources may include occupational incumbents, supervisors, job analysts, or some other type of occupational expert, it seems important that extreme care be taken in both aggregating data to the individual occupational level and in labeling the approaches taken and sources used in the collection of data for occupations.

There are approximately 1,100 occupations anticipated in the O*NET™ database if, as planned, the Standard Occupational Classification forms the basis for the O*NET™ occupational classification scheme. Using the standard of five years for currency of occupational information, it becomes clear that the populating and maintenance of the O*NET™ database is an enormous undertaking. If all occupational information is outdated after five years, then data for about 220 occupations must be collected and integrated annually. Of course, it may be that many occupations will not require updating on this frequency, but others may require more frequent updating. It seems that a systematic method of monitoring changes in occupations to identify "out-of-sequence" needs for updating, as well as a regular schedule for updates, would be essential. All of these activities are both technically and practically feasible given adequate resources.

An essential element is securing the cooperation of establishments and data sources identified as possessing the knowledge to provide data for O*NET™. Usually, a job analysis is carried out for an organization that, for one reason or another, wishes to obtain information immediately. This ensures a fairly high level of cooperation from establishment(s) and people in the organization knowledgeable about the occupation or occupations that are the foci of the analysis. Such is not the case for the O*NET™ database. It is a national database intended for widespread use by a variety of users and, as such, constitutes an extremely valuable resource for many institutions, organizations, and individuals. However, for each individual establishment and data source, randomly or otherwise selected and asked to cooperate in providing data, there is no immediate payoff for offering the access to data sources and time taken to provide information. In a few years, the incentive to cooperate may be the visible utility of the O*NET™ information in a variety of applications useful to schools, government organizations, private-sector businesses, individuals, unions, and others. Unfortunately, that situation does not yet exist. Therefore, it may be necessary to consider a variety of more immediate incentives to encourage cooperation—for example, money, provision of services such as customized data analysis and reports, or symbolic recognition of some sort (e.g., an O*NET™ all-star organization).

## New Technological Horizons

As of this writing, O*NET™ has just been released in its first operational version, called O*NET™ 98. The U.S. Department of Labor anticipates that O*NET™ 98 will enable DOT users to prepare for O*NET™ in the 21st century, when it will formally replace the DOT. The capabilities of O*NET™ as an electronic repository of occupational information, in conjunction with the Internet as a global communications medium, expand the potential of the computer software industry to serve the needs of users of occupational information through the development of highly customized applications interface software. It is also possible, and therefore likely, that the technologies that will facilitate access of the user community to O*NET™ will also be capable of providing information to the Department of Labor (or other O*NET™ maintenance entities) about the uses of O*NET™ data. These technologies could also be used in data gathering to update O*NET™'s databases.

Throughout this volume we have repeatedly emphasized the need for up-to-date occupational information systems that serve the needs of job seekers, career counselors, training specialists, public- and private-sector employers, and state and federal labor and manpower specialists. Occupational information that is broadly encompassing of work as it is performed by a substantial majority of the American workforce, coupled with user-friendly technologies that make the information accessible to users, will facilitate achievement of such objectives as designing new jobs and redesigning existing ones, combining similar jobs or splitting overly complex ones, creating teams and cross-training members, and maintaining systems for staffing, training, and compensation.

This new technical potential may fundamentally alter what is accomplished with occupational information, how it is accomplished, and by whom. Furthermore, technology is likely to keep changing how these systems run and who runs them. For example, making occupational analysis systems available over the Internet may change the way employers recruit and screen candidates, and it may change who does the recruiting and screening. America's Job Bank is an example of an employment service sponsored by a partnership between the Department of Labor and the state-operated Public Employment Service that has been available on the Internet for several years (at www.ajb.dni.us). The number of Internet sites that provide similar services, targeted to particular geographic regions, occupations, and industries, is proliferating rapidly (e.g., the ComputerJobs Store, Inc., sponsors a web site for those seeking jobs in the computer industry at www.computerjobs.com). As we discuss in detail in Chapter 6, similar challenges face the U.S. military as it adapts its occupational structures to new technologies and changing missions. For example, the U.S. Army task lists are 50 percent shorter than in the past, and they are now updated by training school personnel rather than Army Research Institute staff.

The committee anticipates that the availability of O*NET™ will accelerate the development of applications software available from private-sector vendors to serve the needs of the user community. It seems likely that increasing use of O*NET™ databases will increase pressure on the Department of Labor to complete the full database and maintain its currency. It also seems likely that occupational software applications will proliferate, with competitive pressures of the marketplace determining product success. We note that the same phenomenon occurred for products based on the DOT; however, because of the much greater scope of work descriptors of O*NET™ and its electronic medium, we expect that the number and variety of occupational software products available within a few years will be much greater than in the past.

Research and developmental work are needed in both the public and private sectors before these predictions become reality. However, the advent of O*NET™ and the Internet make some version of this scenario virtually inevitable. We conclude with speculations about how O*NET™, when coupled with applications software, could satisfy user needs and, in so doing, contribute to national economic development. Specifically, we provide two brief illustrations of how O*NET™-based occupational information technology could be used to address today's workforce challenges.

Consider the cases of Sal Carpinella and Stan Adamchick, two workers displaced by a defense shipyard in Philadelphia. They found their way into new jobs with the help of career search and job placement software available at the career transition center on base. If Sal and Stan were to face a similar situation again, they could, in the not too distant future, conduct their career searches with their home computers and the Internet.

Sal, who wished to remain an electrician, could request a listing of job openings for electricians in the Philadelphia area. He could also search by O*NET™ occupational code number, or SOC code, to produce a somewhat broader range of options (jobs that require the skills of an electrician but that have a different title would also appear) and could filter the solution further according to employer requirements (e.g., amount of related work experience required, work schedule), job characteristics (e.g., whether the work is performed in a team context, the type of equipment to work on), or other features of work described by the hundreds of work descriptors in O*NET™. Once suitable prospective employers were identified, Sal could electronically transmit his resume to them.

Stan's use of the same Internet occupational database might be quite different than Sal's, due to his desire to change career directions. Stan could explore his options by filtering the database for occupations that match his interests (e.g., thinking creatively) and skills (e.g., troubleshooting problems with electrical systems). He could then investigate the occupations that meet his criteria in great detail, browsing the work descriptors for each occupation. Finally, he could explore the job prospects in his geographic area by searching, as did Sal, for employers with job openings in his chosen field.

Consider another scenario, the case of Tom Johnson, a business unit manager of an auto plant. His company has evolved its vehicle assembly process from a traditional assembly line operation with narrowly defined sequential jobs to team-based assembly with closely coordinated jobs. Workers are now organized into assembly teams and cross-trained to perform a broader array of tasks. This process redesign has paid off in fewer defects per vehicle. Tom is contemplating additional changes by linking technical support staff (helper-mechanics, industrial engineering technicians) more closely to teams, perhaps using a modified matrix model. He is uncertain whether it would be more efficient to add competencies to assembly teams by adding members with desired skills, or to expand the skill sets of team members via training.

Tom and his assistant Jill Turbanski use a software application based on O*NET™ to help answer his questions. They begin by identifying the jobs in the O*NET™ database that correspond most closely to the assembly team and staff support jobs, considering such factors as title, tasks performed, and tools and equipment used. Examination of the broad array of descriptors in the O*NET™ database, however, leads to a discussion between Tom and Jill about which factors are most important to consider in deciding between the alternatives. Jill suggests that a job matching function in the database can be used to explore the similarities and differences between the jobs. Before running the match, they decide to limit the factors included to the specific knowledge, skills, and abilities required for each job. They know that the tasks performed by incumbents of each job are different (hence the idea to add capabilities to the team), so there is no point in comparing jobs on tasks performed. Likewise, there is little need to consider other more general types of work descriptors for this purpose, such as generalized work activities and personality and interest variables, as the issue is work process redesign rather than staffing or vocational counseling. Comparing jobs on knowledge, skill, and ability requirements is the appropriate level of detail, as it addresses the capabilities of the workers to perform the work.

Tom and Jill focus on the top six factors in each category for each job, with factors ranked according to level or amount of each required. In a nutshell, this analysis reveals that the similarities between the assembler and helper-mechanic jobs are much greater than those between the assembler and industrial engineering technician jobs. Assembler and helper jobs require similar amounts of knowledge of mechanics (i.e., machines and tools), engineering and technology (i.e., uses of equipment, tools, and mechanical devices), building and construction (i.e., materials, methods, and appropriate tools), and mathematics (i.e., numbers, their operations and interrelationships). Likewise, they correspond closely on the skill levels required in the areas of operation and control (i.e., controlling operations of equipment), installation (i.e., installing equipment, machines, wiring, or programs), and equipment selection (i.e., determining the kind of tools and equipment needed). Finally, they require similar amounts of ability in manual dexterity (i.e., making coordinated hand movements) and near vision (i.e., seeing details at close range). There is little correspondence between the knowledge, skills, or abilities required of assemblers and industrial engineering technicians.

Tom decides to increase the equipment maintenance and repair skills of the assembler teams by training selected team members. He increases the capabilities of the teams to monitor and optimize their work processes by designating an industrial engineering technician as a part-time member of each team. Before instituting these changes, Jill will gather local data on specific knowledge and skills of both assembler and engineering tech jobs at the plant to supplement O\*NET™ information. These data will be used both as a check on O\*NET™ results and, being more detailed, in designing the equipment maintenance training program for assemblers.

## Final Comment

It may be that O\*NET™ and the SOC will receive adequate human and financial resources and will fulfill the objectives that they have been designed to meet. Even if they do not, however, we believe the nation needs an accurate, current occupational classification system that encompasses both occupational categories (the SOC role) and occupational attributes (the O\*NET™ role) as depicted in Figure 5.1. Individuals and institutions need the information in such a system to plan their futures in the changing world of work, and the development and maintenance of the information in the system is unlikely to be successfully achieved by the private sector. The private sector will no doubt do a superb job of shaping the information resident in the system to meet a wide variety of anticipated and unanticipated consumer needs, but the resources necessary to collect, screen, and integrate occupational data into the system in a timely fashion are properly found in the public sector. A timely and flexible national occupational information system, although a monumental undertaking, is an indispensable public resource and should be supported by public funds.

If these data collection and quality control functions are left to the private sector, it is the committee's opinion that the result will be a patchwork of databases of uneven or unknown quality. Larger, better-financed private-sector organizations or industry sectors may develop occupational information systems to meet their needs, but these systems would, naturally enough, cover only those occupations and descriptive variables that would be of most current interest to them. They may, again understandably, maintain proprietary control over the information to retain a competitive edge. Smaller organizations and sectors may have no systems at all. Trying to put together a national occupational database by building on such a collection of independently developed databases is doomed to failure. Developing and maintaining a common, national occupational database allows private organizations and individuals to apply their resources to augment, enhance, and build on the national information to compete in a global marketplace. Competition is not eliminated, it is likely to be raised to another level—a good thing for the entire country.

One missing element in the present vision of an occupational information system is a closer tie-in of the day-to-day labor transactions and the occupational information system. If daily recruiting, hiring, and firing activity could be linked with occupational categories and, in turn, with the associated skills, abilities, and other attributes of the categories, then trends in desired or required occupations and occupational attributes could be more dynamically monitored. Furthermore, historical data could be accumulated that would be extremely useful for disentangling relatively minor or momentary trends from longer-term shifts in the world of work, something that we have shown to be a difficult undertaking. It seems to us that many of the pieces for such a linkage are already or nearly in place.

To achieve a dynamic, accurate occupational information system, some compromises will no doubt need to be made between breadth of coverage (numbers of occupational categories), depth of coverage (number and types of attributes of occupations), and currency of information (frequency of updating). One possible compromise that strikes us as attractive is to forgo the routine, random sampling of small-frequency occupations or occupations that are extremely difficult to access. The cost per bit of information may be extremely high for such occupations. Such occupations should not be ignored, especially not those that are thought to be emerging or fast-growing occupations, but more directed efforts at data collection that are less costly should be considered. For example, it might be possible to sample from locations where the occupations are known to exist in some numbers and to use matrix sampling of occupational attributes (rather than collecting data on all attributes).

Making predictions about the future is hazardous in any field, and the future of occupations is no exception. However, the existence of an accurate, comprehensive, current occupational database, especially if it is linked to labor transactions as mentioned above, should provide an empirical base for making statistical, algorithmic projections as well as a solid footing for subjective estimates by occupational experts of all stripes.

# 6

# Army Work and Approaches to Occupational Analysis

This chapter focuses on the U.S. Army as a case study in the changing nature of work. It focuses on the application of the concepts, issues, and principles presented in previous chapters. The Army is a unique organization in many ways, set apart from the civilian sector with respect to its function, structure, and place in society. Yet it is also important to recognize that the Army is part of the social fabric of the nation, experiencing in various ways the same trends that affect civilian employers. For example, demographic changes in the general population will ultimately be reflected in the military, along with the forces of change that influence the content and structure of work.

In our analysis, we draw on the framework for conceptualizing the changing nature of work and occupational analysis (Figure 1.1), which has been used as a unifying theme throughout the book. However, before examining the Army in the context of this framework, it is important to recognize the similarities and differences between the Army and employers in the civilian sector.

## KEY FEATURES OF ARMY MISSION AND EMPLOYMENT

Understanding the nature of work in the Army means first understanding the nature of the Army. The U.S. Army mission, as officially stated, is to preserve the peace and security and provide for the defense of the United States; to support national policies; to implement national objectives; and to overcome any nations responsible for aggressive acts that imperil the peace and security of the United States (http://www.army.mil/mission-vision.htm). In much simpler terms, the Army exists primarily to be the nation's warriors and protectors. Both officers and enlisted personnel must be prepared to use force as required and must be willing to stand in the way of prospective violent acts committed by others. The armed services are the only organizations that forthrightly presume that their employees will sacrifice their lives as part of their jobs. Some social scientists suggest that the central skill of military officers is "the management of violence" (Huntington, 1959).

The recent end of the cold war and other geopolitical events have placed Army personnel in new situations and redefined its role with respect to "preserving peace and security." Along with its central and more traditional role of warfighter, the Army and the other services have been asked increasingly to act as peacekeeper and peacemaker, and as an instrument of international humanitarian aid. These added roles are often strange to many military members who may not have received adequate training to deal with the varied aspects of operations other than war. New missions can be sources of profound change since they may call for major modifications in the way people are selected, trained, assigned, evaluated, managed, and used.

Army personnel are currently governed by a highly structured set of rules and regulations. Some of these rules are presented below as a means for drawing distinctions between the Army and the civilian sectors. Although most of the features are fixed at the present time, it may be useful to revisit the rules in light of new missions and personnel mixes, the long-range impact of all-volunteer recruiting, and changing performance requirements.

### Structure

All of the aspects of bureaucratic structure in the Army and other services can be translated into a loss of independent control on the part of the individual members. The details of service are strictly and widely regulated to ensure smooth operation and equity in a very large organization. First, the Army has a fixed-rank structure, determined by federal law, with the number of incumbents fixed at each rank. Individuals who enter the Army generally do so at the very lowest level. Since there is no lateral entry into the organization, as there is in most civilian enterprises, the Army must "grow its own" employees and leaders. Individuals who remain in service form a pipeline to higher grades or levels of authority and experience in occupations.

Second, Army enlisted personnel serve fixed terms of service, although the Army has experienced a first-term attrition rate (that is, recruits who fail to complete their first obligated term of service) of approximately 30 percent since the end of conscription in 1973. Unlike procedures in the civilian sector, employee dissatisfaction cannot usually be translated into an immediate resignation or separation from duty because of contractual requirements. Also, Army personnel are not allowed to organize for the purpose of collective bargaining or for claims regarding working conditions, pay, or benefits.

Third, the pay structure of the Army is fixed. It is entirely determined by rank and time in service, although there is the option of supplementary or special pay and bonuses (such as for reenlistment or service extension of persons in hard-to-fill specialties). A key element of the pay system is that bonuses or merit pay cannot be used in individual cases for rewards or incentives. Merit is recognized through awards (nonmonetary) and medals, whereas superior performance over the longer-term may be rewarded with special honors, early selection (within limits) for promotion to the next higher rank, or selection for a higher position of leadership within the rank pyramid. Military compensation also includes many facets not found in civilian employment, such as pay allowances, benefits, and supplementary pay for support of dependents. In addition, the retirement system requires a minimum of 20 years of service before qualification (unless a special release program authorizes early retirement, as was the case during the force reduction of the 1990s).

Fourth, the conduct of the Army's members is subject to the Uniform Code of Military Justice, along with the U.S. criminal and civil codes. As an earlier report of this committee (National Research Council, 1997a:257) observes:

> In most organizations, a violation of company policy may result in being fired; in the military, violation of company policy may result in formal charges, trial, and imprisonment. . . . And to enforce its standards of conduct, the military maintains its own judicial and penal systems. The mechanisms of control of military members stands in sharp contrast to the treatment of civilians employed by the organization.

In addition, the Army maintains strict control over aspects of its members' personal life, including mandatory drug screening, engagement in fraternization, and sexual conduct, among other areas.

## Functions and Culture

The Army is designed as a community of workers that can be nearly self-sufficient if necessary. A core of soldiers is dedicated to combat activities that are complementary and interdependent. The culture of these fighting units is centered around discipline and a teamwork approach to tasks. They are augmented by large numbers of support personnel who provide everything from administrative services to equipment repair to transportation to meals and housing to medical and dental services to training and all other aspects of organizational need.

All active-duty soldiers are considered to be on a permanent 24-hour call for service, they may be called to duty at any time, including while on leave, and they may be deployed to a duty station far from their home on very short notice. This is a condition of their employment contract and they do not have the right to refuse such duty or to resign. In fact, during Operation Desert Shield and Operation Desert Storm, a "stop loss" action was initiated in order to postpone the discharge of active-duty personnel, regardless of scheduled transitions, expired terms of service, or retirement.

The greatest distinction between the Army and the civilian workforce is one of culture. The structure of military units is hierarchical and their operation is guided by rules of order and discipline. Discipline, in the military context, is more than a requirement to obey the orders of a superior, it also involves following a set of general rules of behavior, even in the face of distractions that include lethal threats. The emphasis on discipline serves two purposes: it facilitates the accomplishment of difficult missions and it supports the creation and maintenance of esprit and morale.

The families of service members are considered to be a part of the Army community. The Army expects certain forms of compliance from family members but also offers a wide scope of support to them. Such support includes religious or pastoral services, social work assistance, family housing, schools for dependents, day care for small children, commissary facilities, and medical care.

The work culture of the civilian economy as contrasted to the Army is less formal and the trend is toward less hierarchical, flatter work structures. Civilian workers have more flexibility in work hours and more choices in terms of the jobs they perform and the organizations for which they work. Even with the military downsizing, however, civilian workers have less job security than their military counterparts.

Although the Army and the other services have continued distinctive organizational practices based on rank and discipline, there are those who believe that changes are coming (Moskos and Wood, 1988; Moskos, 1992; Segal, 1993). These researchers suggest that, because the all-volunteer nature of the force requires competition with the civilian sector for employees, military jobs will become more like civilian jobs in order to appeal to potential recruits. For example, over the years there has been some movement toward making military pay more equal to pay for similar civilian jobs. Also, the Army has become more lenient regarding the justification for attrition during an enlistee's first term.

## Control

The Army, as an organization, is controlled in many important respects by the American people and their elected representatives. Indeed, the Army has been called the "service of the people," because of its size, history, tradition, generally high profile, and previous dependence on "citizen-soldiers" through conscription. One of the most effective controls over the organization is the Army's annual budget, which is approved by Congress each year and affects the authorized number and distribution of personnel. The adoption or elimination of weapon systems can likewise affect the occupational distribution of soldiers, as do congressional decisions regarding the Army's missions, structure, or function. The budget process allows for a level of subcategory specification that targets specific programs or operations; staffing of these operations must frequently be accomplished by diverting personnel from other programs, rather than by adding new personnel to the organization.

### Downsizing Procedures

The Army's post-cold war downsizing is in some respects closer to the personnel cuts in other public or private organizations than it is to armies of previous eras, which demobilized largely conscripted forces after the conclusion of a major war (McCormick, 1998). This relates to the fact that service in the Army has been entirely voluntary since the end of America's involvement in the Vietnam war. The end of conscription and the later conclusion of the cold war have also been identified as seminal events in transforming the very nature of the U.S. armed forces. Since 1987, the Army's active-duty force has been reduced by more than one-third and the budget has decreased by 40 percent.

The reduction of the force called for at the end of the cold war was guided by four principles: protecting quality, shaping the force, maintaining personnel readiness, and demonstrating care and compassion (McCormick, 1998). The need for "force shaping" was driven by the understanding (mentioned above) that the Army would basically be stuck with the results of the reduction-in-force for a long time. Thus, the downsizing was aimed at maintaining the "experience content" of the force, gauged in terms of years in service (Timenes, 1996); a great effort was made to ensure that the loss of personnel would be proportional within all categories of soldiers by their years of service, across certain broad skill areas. Programs and policies for force reduction also attempted to ensure that the vast majority of people leaving the

Army would do so voluntarily. This was achieved primarily through two monetary separation incentives, the special separation bonus (a lump-sum payment) and the variable separation incentive (an annuity). As McCormick (1998) found, the long-term effects of the force reduction are still undetermined, although the Army appears to have been relatively successful in achieving its goals. These procedures illustrate the drive within the Army to make political decisions into rationalized practices.

## EXTERNAL CONTEXTS OF WORK

This section presents a brief discussion of trends in demographics, missions, and technology and their roles in influencing the structure and content of Army occupations. These forces need to be taken into account by occupational analysts in the Army in the same way that similar forces in the civilian sector are considered in the development and use of occupational analysis systems for civilian work.

### Demographic Change

Personnel in the Army and the other military services share certain demographic characteristics that are quite unusual when compared with their counterparts in the civilian workforce. This is due largely to laws and policies that restrict military service to persons who meet specific qualifying standards—including requirements related to age, health, moral background (e.g., arrest records and previous use of drugs or alcohol), physical attributes, marital status and dependents, cognitive ability, gender, citizenship status, education, and sexual orientation, among other personal or background characteristics. However, as the data below show, the Army workforce, like its civilian counterpart, is becoming more diverse.

### General Trends

Since the end of the draft in the mid-1970s, the Army has experienced several noteworthy shifts in its demographic content. First, the proportion of black soldiers in its enlisted ranks has risen

TABLE 6.1  Selected Characteristics of Active-Duty Army Personnel, by Enlisted/Officer Status, September 1996

|  | Percent of Personnel | |
| --- | --- | --- |
| Characteristic | Enlisted | Officer |
| Female | 14.3 | 14.3 |
| Black | 29.9 | 11.2 |
| Hispanic | 6.2 | 3.3 |
| Ages 20-24 | 33.8 | 10.6 |
| Ages 25-29 | 23.3 | 23.4 |
| Ages 40 and above | 5.8 | 24.4 |
| High school graduate (with diploma) or above | 95.8 | 100.0 |
| College degree or above | 10.5[a] | 99.1 |
| Married | 56.0 | 71.4 |
| Pay grade E4-E5 | 45.0 | — |
| Pay grade 03-04 | — | 54.6 |

[a]Includes persons with college credit but no degree, as well as those with a degree.

SOURCE: Defense Manpower Data Center, Monterey, California.

from 14 percent in 1971 to close to 37 percent of new recruits and nearly one-third of all enlisted personnel by 1979. As of 1996 (Table 6.1), the proportion of enlistees who are black had declined by 30 percent, but the relatively high representation of this racial/ ethnic minority group remains as one of the most distinguishing demographic characteristics of the all-volunteer Army (Binkin and Eitelberg, 1982).

Second, in 1973, women comprised about 2 percent of enlisted personnel and 4 percent of officers in the Army. By the late 1980s, this proportion had risen to 11 percent in both the active-duty enlisted and officer ranks; and by 1996, it stood at 14 percent of the Army's enlisted force and 14 percent of the officer corps (U.S. Department of Defense, 1997). Third, in 1973, white men of European descent accounted for about 75 percent of Army active-duty enlisted personnel and 9 in 10 officers; by the mid-1990s, these proportions had declined to 44 percent of the enlisted force and 66 percent of officers (Defense Manpower Data Center, 1998; U.S. Department of Defense, 1997).

Another noteworthy trend has been the increasing age of the Army's active-duty personnel over the past two decades. This has continued throughout the history of the all-volunteer force, as experience levels (number of months or years in service) have risen across the board. The Army's smaller career force has also extended its efforts to recruit enlisted personnel who are some-what older than the traditional 18- or 19-year-old youth fresh out of high school. In 1989, over 62 percent of all new Army recruits were 19 years old or younger, including 33 percent at age 18 and more than 6 percent at age 17. In 1996, just about half (53 percent) of the Army's recruits were 19 years old or younger, with 27 per-cent at age 18 and fewer than 4 percent at age 17. (By law, recruits must be between 17 and 36 years old, and those who are 17 must have parental permission.)

Compared with the other military services, the Army has re-cently tended to take relatively larger proportions of older recruits (over age 21) and lower proportions of younger recruits (17 to 18 years old) (U.S. Department of Defense, 1997). Overall, in 1990, approximately 72 percent of the Army's active-duty enlisted population (men only) were under the age of 30; 6 years later, this proportion had fallen to 66 percent. Some people have referred to this trend as the "maturing" of the Army. Two related conse-quences, one considered positive and the other negative, have been attributed to this so-termed maturing: higher levels of expe-rience tend to translate into improved performance, allowing (in theory) for a trade-off between quality and quantity; at the same time, older soldiers tend to place a greater demand on the person-nel support framework (since they have more dependents) and they cost more (with higher salaries, a greater expense for mov-ing families, and so on).

The demographic trends in the U.S. population described in Chapter 2 and their effects on the general workforce are expected to have both direct and indirect effects on the characteristics of Army personnel. For example, as the proportion of minorities increases in the general population, manpower planners expect to see a corresponding, but smaller, increase of minorities in the Army's ranks; as the mean age of American workers rises, so, too, will the mean age of Army personnel; and, as more women enter the American labor force, most observers anticipate an increasing presence of women in the Army. It should be noted, however, that the Army's workforce is younger and includes a smaller proportion of women than the civilian workforce.

## Sociodemographic Trends

Certain sociodemographic trends in the general population are also expected to affect the Army in ways that are similar to the anticipated impact on civilian employers. Of all the changes that the American family has undergone in recent decades, the Army's personnel planners are perhaps most concerned about the growth of two-income couples, who are expected to represent three-quarters of all married couples by the year 2000. This particular trend has been accompanied by a more equitable sharing of parental responsibilities between men and women—which, along with certain demographic changes, has forced all employers to take greater interest in developing responsive workplace policies and a more family-friendly working environment. More civilian employers are thus implementing flexible work schedules and job-sharing plans, offering expanding opportunities for people to work at home, placing greater emphasis on participatory management, and introducing new compensation packages tailored to the needs of individual workers and their families.

It is clearly more difficult for the Army to create a responsive workplace environment than it is for many civilian employers, given its mission and its corresponding demands on people for their time, availability for deployment, geographic mobility, periodic separations and possible isolation from family, foreign residence, and related obligations of service life. Indeed, both the Army and the family have been called "greedy" institutions, in the sense that each places a great (and often conflicting) demand on the individual for his or her commitment, loyalty, time, and energy (Moskos and Wood, 1988). Yet if the Army is to survive as an all-volunteer organization in a changing demographic landscape, it may have to adapt in a way that can keep step with the movement toward family-friendly work settings.

Recruit quality is another concern of U.S. military manpower planners: that is, whether tomorrow's recruits will have the necessary abilities to perform certain complex tasks in a high-tech

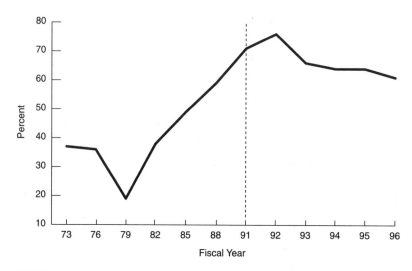

FIGURE 6.1    Percentage of high-quality recruits attracted by the Army during fiscal years 1973-1996.

force. According to Kageff and Laurence (1994:93), military recruits will have to be more versatile: "They will be required to operate and maintain several technically sophisticated systems and, during a course of service, may have to be retrained and transferred laterally." The Army's efforts at attracting high-quality recruits have been quite successful throughout most of the 1980s and early 1990s (Figure 6.1). Applying the Army's own definition of high quality—that is, possession of a high school diploma and a score at or above the 50th percentile (population mean) on the Armed Forces Qualification Test—the trend appears quite positive, although more recently observers have noted test score declines in the military pool of potential recruits (Kageff and Laurence, 1994:92).

One other general demographic trend should be mentioned here. Over the years, considerable concern has centered on the socioeconomic status (SES) of Army personnel relative to that of the U.S. population. When the draft ended, some observers claimed that economic conscription would replace the draft and fill the ranks with disadvantaged young people and that the Army would become an employer of last resort. During the early years of the all-volunteer force, evidence seemed to support the prediction that military recruits would be drawn disproportionately from the nation's lower socioeconomic classes. Since the mid-1980s, however, studies suggest that military recruits tend to come from families in both the middle and lower half of the SES distribution. At the same time, although recruits can be found at all SES levels, persons at the top and the bottom of the scale are underrepresented. These studies, it should be noted, have examined enlisted recruits only and have not taken into account officers, who enter the military as college graduates (U.S. Department of Defense, 1997).

## Occupational Distribution of Women and Racial/Ethnic Groups

Table 6.2 shows the distribution of men and women in the Army's active-duty enlisted force by occupational area. Almost one-third of all male soldiers serve in the Army's general combat

TABLE 6.2 Percentage Distribution of Army Enlisted Personnel, Active Duty, by Occupational Area and Gender, September 1996

| Occupational Area and Code[a] | Male | Female | Total |
|---|---|---|---|
| (0) Infantry, gun crews, and seamanship specialities | 32.9 | 5.8 | 29.0 |
| (1) Electronic equipment repairers | 6.9 | 4.8 | 6.6 |
| (2) Communications and intelligence specialists | 9.9 | 8.1 | 9.6 |
| (3) Medical and dental specialists | 6.1 | 15.9 | 7.5 |
| (4) Other allied specialists | 2.9 | 2.7 | 2.9 |
| (5) Functional support and administration | 13.3 | 36.2 | 16.6 |
| (6) Electrical/mechanical equipment repairers | 14.6 | 8.3 | 13.7 |
| (7) Craftsmen | 2.0 | 1.6 | 1.9 |
| (8) Service and supply handlers | 10.9 | 16.3 | 11.6 |
| (9) Nonoccupational | 0.6 | 0.4 | 0.6 |
| Total[b] | 100.0 | 100.0 | 100.0 |

[a]Based on the Department of Defense (DoD) occupational conversion index. DoD numerical designator appears in parentheses.
[b]Percentages may not add up to 100 due to rounding.

SOURCE: Defense Manpower Data Center, Monterey, California.

TABLE 6.3 Women as a Percent of Army Active-Duty Enlisted Personnel
Assigned to Occupational Areas, September 1996

| Occupational Area and Code[a] | Percent Who Are Women |
|---|---|
| (0) Infantry, gun crews, and seamanship specialities | 2.9 |
| (1) Electronic equipment repairers | 10.4 |
| (2) Communications and intelligence specialists | 12.1 |
| (3) Medical and dental specialists | 30.3 |
| (4) Other allied specialists | 13.3 |
| (5) Functional support and administration | 31.2 |
| (6) Electrical/mechanical equipment repairers | 8.7 |
| (7) Craftsmen | 11.6 |
| (8) Service and supply handlers | 20.1 |
| (9) Nonoccupational | 9.4 |
| All Areas | 14.3 |

[a]Based on the Department of Defense (DoD) occupational conversion index. DoD
numerical designator appears in parentheses.

SOURCE: Defense Manpower Data Center, Monterey, California.

areas (designated as infantry in the table), and somewhat more
than a third of women serve in functional support and administration. In fact, the combination of three areas—functional support and administration, medical/dental specialists, and service
and supply handlers—accounts for over two-thirds of Army enlisted women (compared with less than one-third of their male
counterparts). Also, as Table 6.3 shows, almost one-third of all
personnel in the functional support and administration and the
medical/dental service areas are women. This was more than
twice the level of representation (14 percent) by women in the
enlisted force as a whole in 1996.

A similar pattern of participation by men and women can be
found when looking at the active-duty officer corps. Nearly half
(45 percent) of all male officers serve in tactical operations (the
core warfare areas), and an almost equal proportion (48 percent)
of female officers serve in health care. In fact, almost one-third of
all health care officers are women. Based on the overall percentage of women in the officer corps (14 percent), however, women

are underrepresented only in tactical operations and as general officers—and by a very wide margin.

The reasons for such disparities between the participation of men and women in Army occupational areas can be attributed to legal and policy restrictions on women in combat, tradition, and personal choice. On the matter of personal choice, previous research suggests that women who join the military tend to prefer jobs that are traditionally associated with women (Binkin and Eitelberg, 1986). One factor to keep in mind here is that women, although underrepresented in many areas (based on overall levels of participation) are not unrepresented. As Table 6.4 indicates,

TABLE 6.4 Number and Percentage of Military Occupations with 10 or More Personnel That Have No Women, by Service and Officer/Enlisted Status, 1996

| Service | Officer | Enlisted | Total |
|---|---|---|---|
| Army | | | |
| All occupations | 188 | 250 | 438 |
| No women | 19 | 35 | 54 |
| Percent | 10.2 | 14.0 | 12.4 |
| Navy | | | |
| All occupations | 526 | 999 | 1,525 |
| No women | 104 | 263 | 367 |
| Percent | 19.8 | 26.4 | 24.1 |
| Marine Corps | | | |
| All occupations | 123 | 316 | 439 |
| No women | 49 | 64 | 113 |
| Percent | 39.9 | 20.3 | 25.8 |
| Air Force | | | |
| All occupations | 257 | 243 | 500 |
| No women | 29 | 17 | 46 |
| Percent | 11.3 | 7.0 | 9.2 |
| Total | | | |
| All occupations | 1,094 | 1,808 | 2,902 |
| No women | 201 | 379 | 580 |
| Percent | 18.4 | 21.0 | 20.0 |

SOURCE: Defense Advisory Committee on Women in the Services, *Utilization of Women Indicator Report* (Monterey, CA: Defense Manpower Data Center/Naval Postgraduate School, 30 September 1996), p. 53.

women are missing from only about 12 percent of all Army occupations with 10 or more personnel, a percentage considerably lower than is found in the U.S. Marine Corps, the U.S. Navy, or the U.S. Department of Defense as a whole—even though women are prohibited from serving in ground combat operations, in which 30 percent of Army enlisted personnel can be found. (It is worth noting that only 35 percent of all military enlisted occupations were *open* to women when the nation initiated the all-volunteer force.)

Table 6.5 shows racial/ethnic distributions across enlisted occupational areas. As seen here, whites, Hispanics, and "others" tend to be concentrated in the infantry; blacks tend to be most highly concentrated in functional support and administration. The level of these concentrations can be better observed in Table 6.6, which shows the proportion of all enlistees in each occupa-

TABLE 6.5 Percentage Distribution of Army Enlisted Personnel, Active Duty, by Occupational Area and Racial/Ethnic Group, September 1996

| Occupational Area and Code[a] | White | Black | Hispanic | Other | Total |
|---|---|---|---|---|---|
| (0) Infantry, gun crews, and seamanship specialities | 32.3 | 22.6 | 31.1 | 27.3 | 29.0 |
| (1) Electronic equipment repairers | 6.8 | 6.3 | 6.0 | 6.3 | 6.6 |
| (2) Communications and intelligence specialists | 11.5 | 6.7 | 8.0 | 8.1 | 9.6 |
| (3) Medical and dental specialists | 6.9 | 8.1 | 8.4 | 9.9 | 7.5 |
| (4) Other allied specialists | 3.0 | 2.7 | 2.8 | 3.0 | 2.9 |
| (5) Functional support and administration | 11.0 | 26.3 | 18.6 | 19.0 | 16.6 |
| (6) Electrical/mechanical equipment repairers | 14.8 | 11.4 | 13.1 | 14.4 | 13.7 |
| (7) Craftsmen | 1.9 | 2.0 | 1.7 | 1.9 | 1.9 |
| (8) Service and supply handlers | 11.1 | 13.7 | 9.4 | 9.6 | 11.6 |
| (9) Nonoccupational | 0.8 | 0.2 | 1.0 | 0.4 | 0.6 |
| Total[b] | 100.0 | 100.0 | 100.0 | 100.0 | 100.0 |

[a]Based on the Department of Defense (DoD) occupational conversion index. DoD numerical designator appears in parentheses.

[b]Percentages may not add up to 100 due to rounding.

SOURCE: Defense Manpower Data Center, Monterey, California.

TABLE 6.6  Racial/Ethnic Composition (in Percent) of Army Occupational Areas, Active-Duty Enlisted Personnel, September 1996

| Occupational Area and Code[a] | White | Black | Hispanic | Other | Total[b] |
|---|---|---|---|---|---|
| (0) Infantry, gun crews, and seamanship specialities | 63.9 | 23.3 | 6.7 | 6.1 | 100.0 |
| (1) Electronic equipment repairers | 59.5 | 28.6 | 5.7 | 6.2 | 100.0 |
| (2) Communications and intelligence specialists | 68.7 | 20.7 | 5.2 | 5.5 | 100.0 |
| (3) Medical and dental specialists | 52.3 | 32.2 | 6.9 | 8.5 | 100.0 |
| (4) Other allied specialists | 59.3 | 27.9 | 6.1 | 6.8 | 100.0 |
| (5) Functional support and administration | 38.2 | 47.4 | 6.9 | 7.4 | 100.0 |
| (6) Electrical/mechanical equipment repairers | 62.3 | 24.9 | 5.9 | 6.8 | 100.0 |
| (7) Craftsmen | 57.3 | 31.0 | 5.3 | 6.4 | 100.0 |
| (8) Service and supply handlers | 54.5 | 35.2 | 5.0 | 5.3 | 100.0 |
| (9) Nonoccupational | 74.5 | 10.6 | 10.3 | 4.6 | 100.0 |
| All areas[b] | 57.5 | 29.9 | 6.2 | 6.4 | — |

[a]Based on the Department of Defense (DoD) occupational conversion index. DoD numerical designator appears in parentheses.
[b]Percentages may not add up to 100 due to rounding.

SOURCE: Defense Manpower Data Center, Monterey, California.

tional area according to racial/ethnic group. As the table shows, blacks account for over 47 percent of all active-duty enlisted personnel in functional support; this compares with 38 percent of whites, who comprise 58 percent of enlisted personnel as a whole. Aside from this disparity—as well as in the area of communications and intelligence specialists for whites and blacks—levels of racial/ethnic representation in Army occupations are reasonably close to the proportions found in the entire enlisted force.

As with women, patterns of participation for racial/ethnic groups in the enlisted force are similar to those seen in the officer corps (see Table 6.7). For example, whites tend to be more highly concentrated than other racial/ethnic groups in tactical operations, corresponding to their overrepresentation in the enlisted infantry. The largest proportions of each racial/ethnic minority

TABLE 6.7 Percentage Distribution of Army Commissioned Officers, Active Duty, by Occupational Area and Racial/Ethnic Group, September 1996

| Occupational Area and Code[a] | White | Black | Hispanic | Other | Total[b] |
|---|---|---|---|---|---|
| (1) General officers | 0.5 | 0.4 | 0.04 | 0.1 | 0.5 |
| (2) Tactical operations | 41.3 | 30.4 | 36.3 | 33.4 | 39.5 |
| (3) Intelligence | 6.8 | 5.5 | 8.2 | 6.0 | 6.7 |
| (4) Engineering and maintenance | 9.2 | 13.5 | 10.3 | 10.2 | 9.7 |
| (5) Scientists and professionals | 4.3 | 2.7 | 3.4 | 3.1 | 4.0 |
| (6) Health care | 21.6 | 20.7 | 22.1 | 31.6 | 22.0 |
| (7) Administration | 5.4 | 9.5 | 7.5 | 5.5 | 5.9 |
| (8) Supply, procurement, and allied | 9.2 | 16.3 | 11.3 | 8.9 | 10.1 |
| (9) Nonoccupational | 1.9 | 1.0 | 1.0 | 1.2 | 1.7 |
| All Areas[b] | 100.0 | 100.0 | 100.0 | 100.0 | 100.0 |

[a]Based on the Department of Defense (DoD) occupational conversion index. DoD numerical designator appears in parentheses.
[b]Percentages may not add up to 100 due to rounding.

SOURCE: Defense Manpower Data Center, Monterey, California.

group can similarly be found in tactical operations, but these groups also tend to be somewhat concentrated in engineering and maintenance positions as well as in supply, procurement, and allied specialties. In fact, black officers are overrepresented in supply, administration, and engineering and maintenance occupations. Whites are most overrepresented as general officers, scientists and professionals, and, to a lesser extent, tactical operations officers.

## Migration Between Primary and Duty Occupations

Most studies of military enlisted personnel by occupational category use the service member's primary occupation, or formal area of occupational training, to identify his or her military occupational specialty. Military personnel also have what is known as a "duty occupation," which is the occupation to which they are assigned. Generally, there is a clear connection between a member's duty occupation and his or her primary occupation, or it is assumed that they are the same. We undertook an analysis to determine if the primary occupations of active-duty enlisted personnel matched their duty occupations during the years 1987 and 1997. The results of this analysis appear in Tables 6.8 and 6.9.

Table 6.8 shows the results by broad (one-digit) occupational codes. Almost 16 percent of Army personnel who received infantry-related training as their primary occupational specialty were assigned to some other type of occupation as of 1997. At the same time, 12 percent of personnel trained in functional support and administration were performing some other job; the mismatch of primary and duty occupations for the entire enlisted force that year was 9 percent.

Further investigation focused on primary-duty mismatches

TABLE 6.8  Number and Percent of Army Enlisted Personnel Serving in a Duty Occupation That Does Not Match Their Primary Occupation, by One-Digit Occupational Category, 1987 and 1997

|  | 1987[b] | | 1997[b] | |
| --- | --- | --- | --- | --- |
| Selected Two-Digit Category[a] | Number | Percent | Number | Percent |
| Infantry (0) | 10,414 | 6.8 | 17,219 | 15.9 |
| Electronic equipment repairer (1) | 1,997 | 6.6 | 1,409 | 5.3 |
| Communications and intelligence (2) | 5,042 | 6.4 | 2,017 | 5.2 |
| Health care (3) | 1,141 | 2.7 | 1,453 | 4.5 |
| Other technical and allied specialists (4) | 681 | 4.3 | 612 | 5.0 |
| Functional support and administration (5) | 9,030 | 8.5 | 8,205 | 12.0 |
| Electrical/mechanical equipment repairer (6) | 4,437 | 4.6 | 2,649 | 4.8 |
| Craftsworkers (7) | 774 | 5.6 | 422 | 5.4 |
| Service and supply handlers (8) | 2,592 | 3.8 | 1,345 | 2.8 |
| Nonoccupational (9) | — | — | — | — |
| All Areas | 36,108 | 6.0 | 35,331 | 8.9 |

[a]Based on the Department of Defense (DoD) occupational conversion index. DoD numerical designator appears in parentheses.

[b]As of the end of September of each year.

SOURCE: Derived from special tabulations provided by the Defense Manpower Data Center.

TABLE 6.9 Number and Percent of Army Enlisted Personnel Serving in a Duty Occupation That Does Not Match Their Primary Occupation, by Selected Two-Digit Occupational Category, 1987 and 1997

| Selected One-Digit Category[a] | 1987[b] | | 1997[b] | |
|---|---|---|---|---|
| | Number | Percent | Number | Percent |
| Infantry (01) | 5,514 | 7.6 | 15,204 | 24.8 |
| ADP computer (15) | 92 | 8.4 | 72 | 16.6 |
| Teletype/cryptology equipment (16) | 149 | 6.5 | 59 | 12.2 |
| Radio and radio code (20) | 1,698 | 5.9 | 204 | 14.2 |
| Mapping, surveying, drafting, illustration (41) | 258 | 7.9 | 228 | 13.8 |
| Personnel (50) | 2,512 | 10.5 | 1,394 | 8.9 |
| Administration (51) | 2,553 | 11.0 | 1,041 | 8.8 |
| Clerical/personnel (52) | 206 | 13.8 | 4,073 | 79.9 |
| Artillery/gunnery, rockets, missiles (04) | 2,456 | 5.9 | 1,220 | 5.1 |
| Radio/radar (10) | 1,422 | 7.1 | 1,145 | 5.0 |
| Other functional support (55) | 3,063 | 6.7 | 1,262 | 4.5 |
| Automotive repair (61) | 2,408 | 5.1 | 1,397 | 4.9 |
| All areas | 36,108 | 6.0 | 35,331 | 8.9 |

[a]Based on the Department of Defense (DoD) occupational conversion index. DoD numerical designator appears in parentheses.
[b]As of the end of September of each year.

SOURCE: Derived from special tabulations provided by the Defense Manpower Data Center.

by more refined (two-digit) occupational categories. As Table 6.9 shows, close to one-quarter of personnel (over 15,000 soldiers) with infantry training were serving in a noninfantry occupational category during 1997; this type of mismatch was also found for nearly 80 percent of personnel trained in clerical skills.

It is not clear why these mismatches occur. A study of detailed occupational categories suggested that the primary-duty mismatches were prevalent in occupational categories that are associated with specific weapon systems—perhaps indicating that some migration may have occurred when a weapon system was scaled back or eliminated. Another possible explanation could be that the Army of the late 1990s is providing more flexibility to personnel who may wish to migrate from one occupation to another for the purpose of retention or career advancement. Whatever the case, the study results raise a significant question regarding the accuracy of data on the actual occupations of Army personnel. In this respect, one may likewise question the accuracy or usefulness of current methods to describe the nature of Army occupations. If increasing mismatches are due to lateral movement across occupational specialties, and not to irregularities in the data or to reporting errors, then this would suggest a need for a more refined method of defining occupations. A similar situation exists in the civilian sector, in which workers are more actively moving within and across occupations.

## Mission Change

In Chapter 2 we discussed the impacts of changing markets and global competition on work in the civilian sector. To some, these civilian-sector activities may be considered as responses to threats of corporate well-being. The Army responds to changing global threats; dramatic changes in these threats are producing new Army missions and corresponding visions.

Prior to the end of the cold war the Army was generally guided by a singular threat, which led to clear personnel policies, doctrine, and weapon system assignment. With the end of the cold war, however, there have been fundamental changes in the nature of threats and a shift in missions to include more humanitarian and peacekeeping activities. The core premise of present-day strategic planners is that unpredictability has increased. These planners accept that U.S. military forces must now operate in an environment characterized by a more rapid pace of change than at any time in its history (Paige, 1996a).

In this new post-cold war environment, threats are likely to be more diverse, so that it becomes a challenge to clearly define the threats and develop a strategy for addressing them (White, 1996; Kaminski, 1996). One parameter of threat analysis is the assumption that the total threat picture requires the capability to take on two simultaneous large threats (on the order of Iraq and North Korea) separated by large geographic distance, while deterring any other smaller aggressors from taking advantage of those conflicts (Paige, 1996a; White, 1996).

The proposition that there is a shift from a few large threats to many small threats is confirmed by historical data. For example, during the period 1950-1989, there were 10 major military deployments. Since 1990 there have been 27 major deployments, representing a 16-fold increase, and almost all of them have involved joint operations of U.S. services (U.S. Army, 1997). General John Shalikashvili notes that between Operation Desert Storm and the end of 1996, the military had engaged in over 40 contingency operations.

In addition to expected changes in the types of hostile threats, it is anticipated that military units will engage in numerous and diverse operations other than war, most of which are expected to be highly unpredictable (Shalikashvili, 1996a, 1996b, 1997; Holmes, 1997; National Research Council, 1997a; Osborne, 1997; Association of the United States Army, 1997; Senate Armed Services Committee, 1997). These operations other than war may include:

1. Traditional peacekeeping, which involves stationing neutral, lightly armed troops, with permission of the host state(s), as an interposition force following a cease-fire.
2. Passive observation, which consists of the deployment of neutral, unarmed personnel, with permission of the host state(s), to collect information and monitor activities (e.g., cease fire, human rights, or disarmament).
3. Election supervision, which consists of monitoring polling places and voting procedures.

Other distinctive missions in this category might include humanitarian assistance and disaster relief, pacification of civil disturbances, arms control verification, drug interdiction, and antiterrorist operations.

The conduct of such missions will require military formations of widely variant size and composition. Command and control practices will be particularly challenged and the training of individuals and units will need to be adapted to the nonmilitary dimensions of such missions.

Top military officials at the Defense Department envisage a lean, more mobile, more flexible force with significantly greater ability to rapidly project power or provide assistance anywhere in the world on a moment's notice (Reimer, 1997a, 1997b; Paige, 1996a). Four formal vision statements explain, in general terms, how the armed forces (especially the Army) expect to implement the new military strategies: *The Joint Vision 2010, Concept For Future Joint Operations*, and *Army Vision 2010* documents describe expected requirements for joint operations and Army operations through the year 2010, which is the long-term planning milestone for the military. The Army After Next project description identifies Army requirements projected for the very long-term, from 2010 through 2025.

## Technological Change

Some of the changes in work in the military parallel changes in the civilian sector that result from advances in technology. In both sectors, technology has reduced the number of blue-collar jobs and increased the number of technical and professional jobs. In addition, many jobs in the Army are increasing in scope to include the new technologies. And in some cases, because of shared technology, there is sharing of skills across jobs. This has implications for how work is classified and the training programs that are needed.

Changes in technology interact with other factors, such as: (1) resources (e.g., changing budgets can drive technology and vice versa); (2) political change (e.g., the collapse of the Soviet Union, both affected by technology and producing implications for technological needs); and (3) change in doctrine (e.g., moving from cold war static strategic defense doctrine to one of active defense, which can be facilitated by technology as well as drive it). Moderated by these factors, technology can have implications that reverberate throughout the Army's organizational structure and the content of its work. Although change is a constant in organizations that, like the Army, must adapt to environmental conditions at many levels, technology may accelerate the rate of change, affecting an Army-wide "system of systems" that includes such things as manpower, equipment, training, provisioning, maintenance, evaluation, and doctrine. Box 6.1 provides an example of advances in tank technology.

---

**BOX 6.1**
**Advances in Tank Technology**

The M1 tank (the current model is the M1A2), which replaced the 1970s vintage M60 model, embodied dramatic technological advancements in armor, firepower, ammunition, propulsion, and crew safety. In addition, the large-scale application of the microchip upgraded the tank's systems for fire control, target localization and acquisition, communications, and command and control.

This new technology provided tank crews with greater battlefield situation awareness (Where am I? Where are my friendly forces? Where is the enemy?) and this encouraged further changes in both doctrine and the organizational system for the tank's employment. Armor units could now have greater flexibility on the battlefield, and the scope of execution of order changed. The capability was also introduced to fight with fewer tanks per fighting unit (the platoon and the tank company).

As a consequence, technical training tactics, techniques, and procedures had to be changed for tank crew members, crews, and small units, as well as for all maintenance, transport, and supply personnel connected with the tank. These new task and training requirements led to the need for changes in job classifications and changes in the scoping of individual skills, attributes, and mental ability required to handle the increased complexity associated with the new technology.

---

Advanced technology will be essential to meet the requirements of speed, flexibility, mobility, force protection, smaller staff, information sharing, and awareness of the battle area (Shalikashvili, 1996b). New systems for battlefield use center on digitization but also include other technologies. For example, lightweight armor for both soldiers and vehicles will be more effective in damage protection in the near future. Similarly, the firepower of hand-held weapons will increase. Given the range of battlefield innovations, it will be possible for the individual soldier to be more autonomous while still relying on close teamwork with unit members. For example, fire teams comprised of three soldiers will continue to be the basic infantry combat unit and will constitute a strongly interdependent team. However, the standard doctrine is likely to specify greater spacing so that each team will function more independently of the platoon or company command structure.

Technologies such as the global positioning system will be able to support the fire team leader's new responsibilities for decisions regarding the choice of route for advancing or withdrawing. The actual decision to advance or withdraw will in most instances be made by higher command, but the fire team leader will possess most of the determining information via digitized communications from tactical intelligence sources.

In the future it is expected that new technologies will allow global sharing of information for command, control, and communication functions. Such information is projected to involve high-fidelity images, accurate friend or foe location and identification, smart decision aids, and intelligent filters to reduce, organize, and digest the high volumes of data flowing through the system. It is also anticipated that the future will see new and upgraded applications of information technology for combat support services, such as equipment and system maintenance, logistics, and training. Simplified maintenance and improved reliability of information systems will be achieved by their being self-organizing and self-healing (automated diagnostics and self-reconfiguration). Database management systems that link logistics information (e.g., spares inventory) to maintenance information (e.g., maintenance history and status of equipment) will permit both preventive maintenance and just-in-time delivery of spare parts. Some commanders expect that progressive advancements in technology will produce a metasystem for personnel management. This ideal device would enable commanders to choose soldiers for units, task forces, special team assignments, and duty assignments based on a soldier's proven performance and training on mission-relevant skills and tasks.

Training is another area in which technology is expected to make a substantial contribution. Some examples include the use of high-fidelity simulations for mission rehearsal and the acquisition of new mission-related skills. Embedded, self-paced training that can be tailored to the trainee's level of expertise is also under development.

The Army has developed a program called MANPRINT (Manpower and Personnel Integration) to integrate concerns regarding personnel considerations into the technology development cycle (analysis of requirements, specification, design, testing, and fielding) of newly acquired technological systems. The six areas covered by MANPRINT are: human factors engineering, manpower, personnel, training, health hazards, and system safety. The MANPRINT program applies the following principles for technological procurement (Booher, 1990):

1. User requirements are defined and included in product requirements documents.

2. People considerations are provided in primary decision-making trade-off models.

3. Source selection evaluations weight people factors heavily.

4. Human performance requirements are included in test plans, and tests are conducted with users defined as part of the system.

5. Knowledge, skills, and abilities represented by all human factors disciplines (including organizational analysis and classification) are fully utilized.

6. Capabilities and limitations of users (including operators and maintainers) are taken into account during the definition of product requirements and specifications.

### Organizational Change

As for many organizations in the civilian sector, downsizing has significantly affected the Army's work—in the past decade, the active Army has been reduced by 275,000 (36 percent reduction), the Army National Guard by 84,000 (18 percent reduction), the Army Reserve by 89,000 (28 percent reduction), and the civilian workforce by 135,000 (33 percent reduction). Figure 6.2 shows the progression in the Army's planned downsizing from 1989 to 1997.

The active Army provides a forward presence and an initial rapid response to emergencies; the Reserves and the National Guard provide a pool of trained individuals for active duty in time of war or other emergencies that can act as reinforcements for contingency operations. In today's environment, the two reserve components are being exercised and deployed to a greater

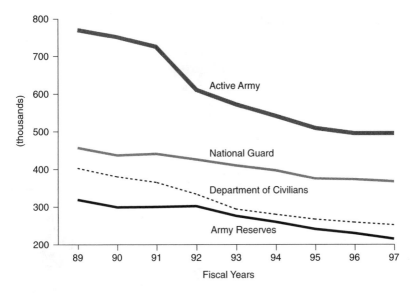

FIGURE 6.2    The Army personnel drawdown, fiscal 1989-1997.

degree than in the cold war period.  The result is increased strain on all components and a requirement for greater flexibility in mission adaptation for both individual soldiers and units.

The total Army is composed of both tactical and nontactical units.  The tactical units include combat, combat support, and combat service support; the nontactical units perform such functions as recruiting, training, personnel management, and materiel command.  The total Army composition in 1997 for tactical units was: National Guard (45 percent), active Army (38 percent), and Army Reserve (17 percent) (U.S. Army War College, 1997).  A further breakdown in shown in Figure 6.3; it shows that the National Guard provides 55 percent of the combat troops, 33 percent of the combat support units, and 26 percent of the combat service support units.  Currently there are 8 combat divisions and 15 separate brigades in the National Guard.  The Army Reserve, in contrast, represents only 3 percent of combat troops, while providing 31 percent of combat support and 46 percent of combat service support units.  The Reserve also provides 100 percent of the railway units and enemy prisoner-of-war brigades, 86 percent of the

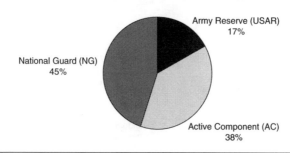

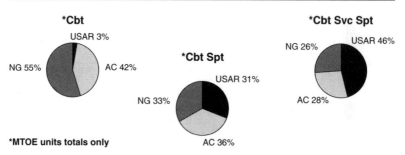

FIGURE 6.3   Total Army composition by component (MTOE only), fiscal 1997.

psychological operations units, and 70 percent of the medical and chemical capability.

To enter the National Guard and the Army Reserves, there are several paths (Figure 6.4). Some of those entering as reserves are individuals who have left the active force sometime during their careers; others, after being recruited, go directly into the reserve forces with no prior military service. According to a recent study (Grissmer et al., 1994), the percentage of individuals in the Army Reserves and the National Guard with no prior service experience was approximately 60 percent for the enlisted force; for commissioned officers, these percentages were 45 for the Army Reserve and 60 percent for the National Guard. Current procedures require that individuals entering the reserves directly must undergo active Army basic training. Generally, this training is accomplished in two eight-week segments called split-entry training. As service in the Army Reserves continues, more specialized

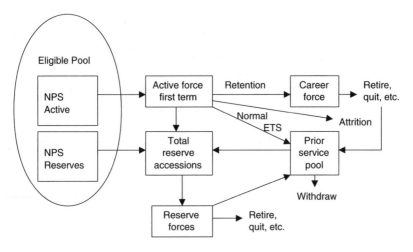

FIGURE 6.4    Total force supply and demand.  (ETS = expiration of term of service)

training is provided.  Both the National Guard and the Army Reserves receive 39 days of training a year, far less than the training of those on active duty.  There is some concern that, as the active force continues to downsize, there will be an even greater proportion of reservists with no prior service, further reducing combat readiness.

## THE ARMY'S APPROACH TO OCCUPATIONAL ANALYSIS

The work structure in the Army provides the basis for selecting, training, organizing, and managing personnel to meet mission requirements.  The result of changing mission requirements has been the development of a smaller, more flexible force with a wider range of fighting skills—as well as new skills in negotiation and interpersonal interaction.  Some of the implications of these changes parallel changes in the civilian sector.  For example, the increased diversity of Army missions coupled with downsizing has led to the creation of teams of individuals with different skills from different work cultures.  Instead of an infantry unit working alone in a combat role, a tradition in the Army, there is

now a need to combine combat troops with Army Reserve nego-
tiators and civilian technicians to perform humanitarian and
peacekeeping missions. As a result, there is an increasing requirement for communication and coordination, both of which may
involve some increase in the emotional labor component of the
work. Also, because of the large number of potential threats
around the world, there is a need to rapidly deploy teams of
people who have not had the opportunity to develop cohesive
relationships.

With fewer personnel, individual soldiers may need a broader
range of skills than in the past, or they may have to work with a
more diverse group of individuals from various parts of the military. The work structure must be flexible enough to adapt to these
changes and must facilitate a rapid response to a wide variety of
situations. According to Sellman (1995:10):

> We can no longer afford the luxury of occupational specialization.
> Personnel again will be trained in broad skills and knowledge that
> are necessary to perform a wide variety of job activities, rather than
> just those skills and knowledge needed to perform a small set of
> tasks. Obviously, such a change in training philosophy will have
> major implications for occupational analysis.

The central question is: Do the current structures and the analysis tools used to design jobs and training programs support the
present and future needs of the Army?

## Occupational Classification in the Army

The personnel of the active Army, the Army National Guard,
and the Army Reserves are all classified and managed according
to a common set of occupational categories. All soldiers in the
Army are assigned to and trained in one of the combat, combat
support, or combat service support branches, depending on the
functions they will perform either in combat or in support of combat.

The occupational structure that is used to classify soldiers is
hierarchical. The enlisted force is divided into 31 career management fields that relate to the Army's branches (Box 6.2). Each
career management field, such as infantry, special forces, and

transportation, includes a series of military occupational specialties (MOSs).

There are approximately 350 MOSs. Every MOS has a promotion structure and a description of the tasks, skill qualifications, training and time in rank requirements associated with every enlisted rank. Enlisted personnel are assigned to an MOS when they enter the Army. Assignment is based on both mental and physical performance standards and on position availability. Mental abilities are measured by the Armed Forces Vocational Aptitude Test Battery (ASVAB). Different MOSs have different standards for assignment. Currently, over 95 percent of recruits are high school graduates and have ASVAB scores in the upper 50 percent. There is no lateral entry for enlisted soldiers. Essentially, the Army recruits potential and provides the training necessary for the individual to perform his or her job assignments. That is, recruits come with aptitudes and the basic capacity to learn the needed skills; the Army does not look for individuals who already possess such skills.

As a trainee enters the infantry, he is assigned to one of four career paths—infantryman, indirect fire infantryman, heavy artillery/armor infantryman, or fighting vehicle infantryman. The levels represent steps in a career progression (from private/specialist to sergeant, to staff sergeant, and so on). At each level, there are specific requirements for training, skill level, and time in service.

Approximately 80 percent of enlisted soldiers stay within the same MOS throughout their Army career. In some fields, however, there are specified points at which a soldier may cross from one MOS to another. Also, personnel may transfer out of a combat arms MOS later in their careers. The system is based on the premise that enlisted soldiers who choose an Army career and move up through the ranks need the training and experience provided at the lower levels of the specialty in order to have sufficient knowledge to take on each higher level of tasks. Although there may be MOSs with overlap in tasks or skills, there is no easy way to efficiently make these links using the current system.

New MOSs may be added and existing ones can be modified on a yearly cycle. A given MOS can be combined with other MOSs or eliminated entirely as a function of new mission requirements

## BOX 6.2
## Career Management Fields

Enlisted Career Management Fields

11 Infantry
12 Combat engineering
13 Field artillery
14 Air defense artillery
18 Special forces
19 Armor
25 Visual information
31 Signal operations
33 Electronic warfare/
   intercept systems
   management
35 Electronic maintenance
   and calibration
37 Psychological opera-
   tions
38 Civil affairs
46 Public affairs
51 General engineering
54 Chemical
55 Ammunition

63 Mechanical mainte-
   nance
67 Aircraft maintenance
71 Administration
74 Information systems
   operations
77 Petroleum and water
79 Recruiting and reten-
   tion
81 Topographic engineer-
   ing
88 Transportation
91 Medical
92 Supply and services
93 Aviation operations
95 Military police
96 Military intelligence
97 Bands
98 Signals intelligence/
   electronic warfare op-
   erations

Office Career Branches

11 Infantry
12 Armor
13 Field artillery
14 Air defense artillery
15 Aviation
18 Special forces
21 Corps of Engineers
25 Signal corps
31 Military police
35 Military intelligence
38 Civil affairs
42 Adjutant General's
   corps
44 Finance corps

55 Judge Advocate
   General's corps
56 Chaplain
60-62 Medical corps
63 Dental corps
64 Veterinary corps
65 Army medical specialist
   corps
66 Army nurse corps
67 Medical service corps
74 Chemical
88 Transportation corps
91 Ordnance
92 Quartermaster corps

Other Functional Areas

35 Military intelligence
39 Psychological operations and civil affairs
41 Personnel programs management
45 Comptroller
46 Public affairs
47 USMA permanent faculty
48 Foreign area officer
49 Operations research/systems analysis

51 Research, development and acquisition
52 Nuclear research and operations
53 Systems automation
54 Operations, plans and training
90 Logistics
97 Contracting and industrial management corps

Medical Functional Areas

70 Health services
71 Laboratory sciences

72 Preventive medicine sciences
73 Behavioral sciences
75 Veterinary services

or technological advances. These changes generally are the result of requests from the field. When changes occur, a consequence is that some soldiers must be reassigned, retrained, or assisted in finding a job outside the Army. According to a cross-walk using the U.S. Department of Labor's *Dictionary of Occupational Titles*, more than 75 percent of military jobs have one or more counterparts in the civilian economy.

The career management fields vary in size—the three largest are infantry (49,837), mechanical maintenance (39,474), and supplies and services (38,526); the three smallest are psychological operations (458), public affairs (688), and topographic engineering (832).

All officers are commissioned in one of the basic branches of the Army. Functional areas are assigned based on the Army's needs and the qualifications of the officer. Most officers are developed within a branch; branch transfers are considered in the third to the eighth year from branches with overstrength to those with understrength. Each position is characterized by the occupational skills required to perform its principal duties; secondary skills and areas of expertise are also listed as part of the overall position description.

Requirements for personnel and equipment are driven by the Army's current and future mission requirements (U.S. Army War College, 1997). Once these requirements are defined, the force (including National Guard and Reserve units) can be configured and sized to best meet the requirements. Manpower managers, working within force structure requirements, determine the number of personnel spaces to be allocated to each grade and skill at both the officer and the enlisted levels. Information on the exact composition of the total force at any given time is provided by a computerized system that accounts for individuals who are operationally available plus those who are in training, hospitalized, traveling, or otherwise unavailable. Personnel managers can maintain readiness by efficiently filling the allocated spaces with qualified individuals listed in the computer system. The Army's occupational structure is at the core of all of these activities.

At the enlisted level, for each MOS, there is a specification of the number of spaces for each rank and skill by geographic location; these specifications change as force structure requirements change. There are also historical data on the amount of turnover at each level and thus the expected vacancies needed to be filled at any one time. These data are used by personnel managers to make decisions about recruiting, training, promotion, and assignment to units. Units are prioritized in accordance with readiness status; those with the highest status are kept at full strength.

Recruiting targets are based on the number of first-term spaces that need to be filled in each MOS, the expected attrition rate during the first enlistment term, and the number of soldiers needed to reenlist to fill the required number of second-term positions. Even with the high quality of recruits, the attrition rate during the first term remains around 35 percent (U.S. General Accounting Office, 1998). These enlistees are separated from active duty before the end of their first term for medical or physical reasons or because they claim special conditions that may legally allow early discharge. In order to fill gaps that are generated by attrition, financial incentives are used to attract recruits to serve in understrength MOSs.

Soldiers entering a second term are generally at the level of corporal (E-4) or sergeant (E-5). Once a soldier reenlists for a second term, he or she is tracked by the U.S. Total Army Personnel Command, providing information on rank, time in rank, training, special skills, location, and time at location. These files are used in conjunction with information on spaces to be filled at each level, in each MOS, and at each geographic location to make decisions about promotion, training, and distribution of enlisted personnel. As soldiers continue their careers and meet criteria for promotion, they are placed on the appropriate promotion lists. Because there may be more individuals eligible for promotion than there are available slots, those on the list are ranked (according to a point system) for purposes of selection. The promotion process at the lower levels (E-1 to E-3) is decentralized, at the midrange (E-4 to E-7) it is partially centralized, and at the top it is centralized (E-8 to E-9). Critical aspects of this progression through the ranks are both skill training and leadership training. The U.S. Army Total Personnel Command maintains several data systems that provide the information needed to manage personnel throughout the Army.

As noted above, most officers stay for their entire careers in the branch they select when entering the Army. Changes are generally made when the structure changes and branches are downsized or when an officer in an overstrength branch applies for duty in an understrength branch. Officers progress through a promotion structure that generally requires time in rank as well as specific training and operational experiences. Above the level of first lieutenant, the promotion process is centralized. Adjustments in the size of the enlisted and officer force and the ratio of officers to enlisted personnel are made as budget and mission requirements change. There is a continuing need to reassess personnel distributions in order to provide the appropriate levels and types of skills in a sufficient number of troops to ensure readiness to meet projected missions.

## Challenges Facing Occupational Analysts

The increasing emphasis on joint operations, particularly when hybrid units are formed (i.e., when personnel from different services combine into low-level units, as opposed to more traditional joint operations strategies that assign specialized tasks to homogenous units within the different services), will require a fundamental change in the Army's culture, and perhaps a generation to fully implement (Depuy, 1995). Joint operations may also eventually involve increased competition between the services, especially if duplicate capabilities are integrated. For example, the United States currently has four air forces, two infantries, three medical corps, and many other units that are duplicated across the services. A major implication is that the different services may be required to develop a common work structure that supports the development and assignment of hybrid teams. For example, a new career opportunity has been created for officers who wish to take a Joint Specialist path (Depuy, 1995).

Rapid expansion (response and on-demand increases in both the size and range of force capabilities) and tailoring (matching force capabilities to mission needs) will be required to respond to sudden, nonstandard events. Thus, the work structure itself becomes an increasingly important tool—a part of the information technology that must be used to speedily identify, locate, and assign large numbers of personnel for specific missions. If these missions are joint, this increases the desirability of a work structure that describes jobs, people, and missions in a common language. Since changes are expected to be ongoing and rapid, the work structure must be continually and rapidly updated; this implies value for automated on-line services with Internet-like availability across the services. Since this work structure and personnel information becomes part of the battle, it must be protected.

New missions—particularly those that extensively or primarily involve interaction with civilians—will require that the work structure identify new tasks; team mixes; and knowledge, skills and abilities. New tasks for operations other than war involve the development of social skills for dealing with civilians and their leaders in such roles as peacekeeping and community service. These tasks also may require special knowledge, skills, and abilities, such as fluency in the native language; cross-cultural sensitivity and knowledge; emotional work skills; and ability to engage enemies who do not follow traditional rules of war (Peters, 1994; Russell et al., 1995; *Force XXI Operations*, 1994; Schoomaker, 1998). Work structures must include descriptors for these tasks and knowledge, skills, and abilities (National Research Council, 1997a).

Extremely high tempo and a strategy that allows attack from multiple positions simultaneously will require commanders who possess suitable decision-making, teamwork, and problem-solving skills, and work structures must include descriptors for these skills. Commanders will be supported by different organizational structures that delegate responsibilities and authorities to lower ranks, and by technologies that permit such delegation by rapid collection, transmission, and analysis of information. Commanders must be able to plan, decide, and execute commands extremely rapidly and with great flexibility—rapidly grasping changes in situations and exercising judgment while reasoning under uncertainty (*Force XXI Operations*, 1994).

Stress response and soldier adaptability may be very important descriptors for all officers and enlisted personnel (Joint Chiefs of Staff, 1997; U.S. Army Training and Doctrine Command, 1996; Reimer, 1997c). The document for the AAN project, which identifies Army battlefield requirements for the years 2010 through 2025, also suggests that the following knowledge, skills, and abilities will be important for all future missions: mental agility under severe circumstances, the ability to translate the commander's intent into effective tactics with minimal supervision, the ability to handle complex tasks, and interpersonal skills such as trust, confidence, and loyalty despite less face-to-face contact.

Finally, a major, debated question with significant potential implications for the Army work structure is: Should the number and type of special operations forces be increased to address the expanded set of unusual missions, or should special forces type tasks be considered part of most soldiers' work (Holmes, 1996; Russell et al., 1994; Schoomaker, 1998)? On one hand, nonstandard operations have traditionally been addressed by special forces who are supported by a training and work assignment structure

geared toward rapid response and tailored personnel and equipment mixes. On the other hand, nonstandard operations—particularly operations other than war that require less "warrior" and
more "warrior-diplomat" skills—may become the norm for the
future Army. The *Force XXI Operations* document (1994) identifies Army battlefield requirements through the year 2010 and suggests that the smaller force will have fewer individual specialties
for both officers and enlisted soldiers, and individual soldiers will
be trained for a wider variety of missions. That would argue for a
thorough reconsideration of the existing work structure in light
of the requirements for these nonstandard missions.

As discussed in previous chapters, changes in technology and
organizational context have contributed to two challenges that
face occupational analysts in the civilian sector: the need to define and respond to the higher skill requirements of jobs, and to
the blurring of distinctions between jobs. These two challenges
are also salient for the Army.

### Higher Skill Requirements (Upskilling)

The mix of Army jobs in the post-World War II period shows
a shift from blue-collar to white-collar jobs, reflecting the move
away from work requiring general military skills toward that requiring special skills (Binkin, 1994). The most conspicuous
change has been the increase in electronics-related occupations.
It should be noted, too, that the growth in the proportion of technical jobs has been accompanied by an increase in the technical
complexity of specific jobs. These trends are closely parallel to
those in the civilian sector. As the armed services draw down
their forces, the requirements for bright, technologically literate
personnel are not likely to diminish proportionately. It is more
likely that the requisite personnel qualifications in the leaner military of the future will grow commensurate with the sophistication of the systems that are fielded.

Paige described the skills associated with using and maintaining sophisticated equipment as follows (1996b:5):

> Let me give you a vivid example of our world of changing technol-
> ogy. Today 82 percent of what (a particular person) maintains is
> computer controlled. He averages 100 hours of training a year. His

average age is 36. Twenty-seven percent of his peers attend college. He deciphers 500,000 pages of technical manuals. The best and brightest who are skilled in computer diagnostics can command $75,000 per year. Does this technician maintain the new Comanche helicopter with its four onboard supercomputers? No. The technician I have described maintains your new automobile. If this is what today's mechanic(s) need to do their job, think what tomorrow's warfighter will need.

A related question is whether "smart" systems of the future will permit use and maintenance by smaller staffs of people. Binkin (1994) suggests that the number of personnel needed by the armed services depends on many factors, such as what tasks specific units are expected to do, how they are organized (combat-to-support ratio), what skills are required, and guiding personnel policies (how people are assigned and used). The influence of technology comes into play when calculating the number of people needed and the qualifications that specialists and technicians need to operate and maintain the military equipment. As new systems and advanced technologies are introduced, the effects on the military work force will largely depend on the degree of equipment complexity, which is directly related to its reliability, maintainability, and availability. Binkin concludes, for example, that the military has consistently been underestimating the number of maintenance personnel that will be needed for new equipment.

The Army's current job classification system distinguishes between electronic equipment repairers and electrical/mechanical equipment repairers; there are currently twice as many of the latter than the former. However this level of distinction is far too coarse. Some truly advanced systems, for example, are likely to require maintenance MOSs that are system-specific. These new MOSs will need to be supported by analyses of tasks, knowledge, skills, and abilities—especially cognitive tasks (National Research Council, 1997b). Different jobs using similar equipment can involve different tasks and skills. By the same token, apparently dissimilar tasks can involve similar fundamental jobs and skills. Only comprehensive task analyses can decompose the nuances of such jobs.

In any case, most military leaders caution against assumptions that future "smart" technologies will replace human soldiers' and commanders' abilities to plan and to make decisions. Although there are clear cases in which complete automation of human tasks is a goal (e.g., mine clearance operations, surveillance under extremely dangerous conditions), improved technologies are intended to augment and assist, not replace, human control of the battle area. Technology will not negate the requirement for basic human warrior skills (Joint Chiefs of Staff, 1997; Holmes, 1997). Military experts endorse the notion that future technology will require not only continued employment of human warfighters, but also higher levels of skill (Sanders, 1997:3).

In addition, the Army's future work structure will have to contend with the fact that workers who deal with information technology must keep up with a staggering rate of growth in the relevant knowledge base. One study found that the half-life for material covered in college engineering classes is between 5 and 10 years, and engineering reference library documents have an even shorter useful half-life of only 1 to 2 years. As a result, workers in information technology careers must constantly enhance their knowledge just to keep up, and today's graduates must be prepared to change careers many times in response to technology changes. The same considerations will apply to military workers (Kaminski, 1996).

## Blurring of Distinctions Between Jobs

Distinctions between jobs may be increasingly blurred due to changes in technology-based skills and new mission requirements. In at least two cases, previously distinct jobs may be at least partially melded in the future. One pair of job types that is losing its distinction is operational users (e.g., tank crews, helicopter crews) and maintainers. The modular design and self-test capabilities of advanced technologies will increasingly permit "pull and replace the box" maintenance, which requires no special tools and can be performed in the field by operational users. In addition, equipment and systems are becoming more software-intensive, and in the future it is possible that operational personnel may be trained to monitor, diagnose, and reconfigure their systems through software control—a significant increase in maintenance responsibilities.

Distinctions are also being lost between commanders and subordinates. Since new information distribution and processing technologies will permit (and strategies will require) commanders to perform their monitoring, planning, decision-making, and controlling functions through interaction with common computer-based systems, they will have to be proficient in the technical skills required to interact with and understand the capabilities and limitations of those systems. That is, command tasks will require more technical skills and knowledge. The same strategy and technologies will permit and require subordinate soldiers to operate at a distance from the commander and to assume more responsibility and authority for tactical actions based on immediate situational feedback and awareness. Osborne observes (1997:14): "The evolution of smart computer terminals has unleashed the potential of the individual to control information flows. The armed forces have already made great use of this process by empowering individual soldiers to make decisions on the battlefield. The new [organizational] structures often require supervisory and nonsupervisory personnel, trained to function in diverse capacities, to cross the lines of conventional job descriptions. That is, a subordinate's tasks will often require some command skills and knowledge." However, as noted earlier in this chapter, rules regarding recruitment, compensation, and status difference remain throughout the enlisted and officer ranks.

### New Organizational Structures and Processes

In the civilian sector, the conventional management hierarchical pyramid is being flattened to provide faster information flow horizontally and from the top down and from the bottom up (Osborne, 1997). Reducing the number of organizational levels promotes teamwork, speeds product development, and allows flexible, rapid response to market changes. Although the military's new strategic emphasis on speed and flexibility of response has implications for a wider distribution of information and work responsibilities, there is little serious consideration being given to flatter, nonhierarchical organizational structures. The Force XXI Operations document (1994) suggests that, in the future, physically dispersed Army organizations are likely to be electronically linked and structured to provide the most timely information directly to soldiers so that they can exercise their full potential for initiative and action within the overall intent of the commanders.

In any case, the military assumes that, in the future, warfare will require adjustments to organizational structures that take advantage of, and may even be organized around, the processes and systems for information processing and distribution (Joint Chiefs of Staff, 1997; *Force XXI Operations*, 1994). Such changes may include changes in the nature of command authority (e.g., authority corresponding to possession of knowledge—and therefore changeable—rather than purely on the basis of rank), are likely to diffuse authority, and will change the dynamics of leader-to-led in ways that are yet to be fully explored and exploited (*Force XXI Operations*, 1994).

Such explorations are likely to benefit from investigation of new work structures being developed in association with the concept of computer-supported cooperative work, in which teams work together using common information-processing and distribution systems, while individual tasks are allocated dynamically based on overall team actions, performance, and requirements (National Research Council, 1998). Army work structures in the future are likely to benefit from inclusion of detailed descriptors for team tasks and knowledge, skills, and abilities.

### Developing an Effective Occupational Information System

In 1965, the Army recognized the need to make use of occupational analysis methodology for creating, revising, and merging occupational specialties. The first effort in this direction was the development of the Military Occupational Data Bank, which later evolved into the Army Occupational Analysis Survey. In 1972, the Army abandoned these systems for the Comprehensive Occupational Data Analysis Program (CODAP), developed by the Air Force.

CODAP is based on the assumption that occupational analysis begins by defining all jobs of interest down to the task performance level. In this approach, the list of specific job tasks is the primary anchor for job data; this list can be augmented or modified by other factors, such as equipment used. Once the task list is completed and appropriate rating scales are added, this preliminary job inventory is reviewed for accuracy and completeness by experienced job incumbents, supervisors, trainers, or other subject-matter experts. The next step is to administer the inventory to a large number of job incumbents to collect quantitative data on the time spent and the importance of tasks (in some cases, other data are also collected). These data are then analyzed, interpreted, and used for a number of manpower management purposes, including especially the development of training programs and the definition of career paths with specific details about the increases in responsibility at each level of advancement. This approach, although labor-intensive and time-consuming, provides a common framework for commanders, personnel managers, and trainers.

In an effort to streamline the occupational analysis process, the responsibility for the system was transferred to the Army Research Institute in 1994. The three biggest concerns of the users of the system at that time were (1) the need to shorten the time needed to obtain occupational data once a requirement was identified, (2) the need for a central on-line database to facilitate analysis, and (3) the need for timely analysis and easy interpretation and use of results.

In 1996, the Army Research Institute's occupational analysis group stated its goal (U.S. Army Research Institute for the Behavioral and Social Sciences, 1996:00):

> The mission of the Occupational Analysis Program is to provide the Army's manpower, personnel, and training communities the individual task information critical to job design, analysis, and training development. It is through the integration of the requirements of these three communities at the military occupational specialty or job level, that the OA Program supports the field commander, the ultimate customer for occupational analysis in the Army.

Its definition of occupational analysis encompasses all aspects of work organization, performance, and training. Job analysis, which is central to creating an effective fighting force, is a critical subset of occupational analysis. In its Army application, it focuses on defining MOSs through detailed description of both the tasks and the knowledge and skills required for effective performance.

Recent efforts to modernize the job analysis process and make it more efficient led to the development of a new computer-based survey system called Operational Data, Analysis, and Structure (ODARS). By combining psychological and computer methodologies, this system offers (1) automated surveys, (2) continuous data collection, (3) a centralized and accessible occupational analysis data base, and (4) flexible, easy analysis and report generation. ODARS has been used with some success in characterizing changing task requirements in selected MOSs and for developing responsive and targeted training programs (U.S. Army Research Institute for the Behavioral and Social Sciences, 1996).

The greatest use of occupational analysis is in the training function; task and knowledge lists are regularly reviewed to ensure that training is up to date. Training courses now are on a 6- to 8-year revision schedule. Experienced job incumbents and supervisors provide task ratings and updating. The goal is to obtain ratings by 100 percent of soldiers in small MOSs, and by every trainee attending a training school in one year for larger MOSs. There is limited capacity to validate task lists; it is assumed that raters are qualified subject-matter experts. Training developers analyze task lists and decide when, where, and how to train. On average, approximately 250 development hours are devoted to every instructional hour per program.

Consolidation of MOSs is an important current issue. A redesign is instigated by a "proponent office" at one of the 26 training schools, in response to one or more sources of pressure to change. These pressures occur due to changes in command, doctrine, technology, quality of human resources, performance problems, downsizing, and so on. Consolidation of MOSs is done on the basis of common knowledge, not common tasks, because MOSs are designed to have nonoverlapping tasks. There are no set procedures on how to design or redesign an MOS, and proponents have no formal training in MOS design. The recent drawdown in military personnel can have dramatic effects on this process due to attrition of professional expertise.

The occupational analysis staff has identified three programs for future development. The first will focus on creating approaches to MOS analysis and job design that will provide a database of common tasks, skills, and knowledge across MOSs. It is suggested that such a database could be extremely valuable for designing new jobs or combining existing jobs. The second development concerns evaluating the applicability and usefulness of various civilian job classification systems, such as O*NET™, both for relevant methodology and for matching military and civilian jobs. The third development involves exploring the use of the Internet for data collection and information dissemination.

In a report prepared for the Army Research Institute, Russell, Mumford, and Peterson (1995) proposed the application of O*NET™ to occupational analysis in the Army. They begin their discussion with a statement of the role of occupational analysis in the Army of 2010 with regard to changing missions, tailoring units to missions, developing new technology and weapons systems, changing battle command, emerging information technology, and ongoing rapid change. Table 6.10 (taken from this report) provides an overview of the implications of these changes for manpower, personnel, and training as well as for occupational analysis.

The report further proposes an ideal Army occupational analysis system that would "be used by manpower, personnel and training professionals and perhaps Army commanders in the Army 2010. Its linked data bases would allow easy access to descriptions of training courses that teach a particular skill, to lists of soldiers who have skills and abilities relevant to a particular type of mission, to Army jobs that have similar requirements, and so on. It would have a menu-driven, user-oriented interface that allows users to access data at the level of aggregation and specificity that is best suited to the application" (Russell et al., 1995:11).

The three linked databases—readiness, occupations, and training—would be related to each other through a common language provided by O*NET™ variables. Together, these databases would provide all the information needed for such tasks as assembling a special operation in the field or for developing training requirements for a combined MOS.

The specific characteristics of an ideal Army Personnel Network (AP*NET) and their relationship to existing O*NET™ char-

TABLE 6.10 Implications of Anticipated Changes in the Army for Manpower Personnel and Training and Occupational Analysis Systems

| Anticipated Change | Implications for Manpower Personnel and Training Systems | Implications for Occupational Analysis |
|---|---|---|
| Changing missions | Must address interpersonal, cross-cultural, and other non-technical knowledges, skills, and abilities in selection and training. | Include descriptors for interpersonal and other non-technical knowledges, skills, and abilities. |
| Tailoring units to missions | Must provide information for rapid team formation. | Describe jobs, people, and missions in a common language. |
| Developing new technology and weapons systems | Must enhance transfer of training across jobs and specific pieces of equipment.<br>Must select soldiers who are adaptable. | Include descriptors of broad technological skills. |
| Changing battle command | Must ensure that soldiers have needed decision making, teamwork, and problem-solving skills. | Include descriptors for decision making, teamwork, and problem-solving skills. |
| Emerging information technology | Must realize that manpower personnel and training information can become a part of the battle. | Include descriptors useful to commanders in the database.<br>Develop policies and controls for use. |
| Ongoing rapid change | Must be continually updated and accessible. | Take advantage of automation and on-line services.<br>Develop future-oriented job analysis approach. |

SOURCE: Russell et al. (1995).

TABLE 6.11  Mapping of Desirable AP*NET Characteristics Against O*NET™ Characteristics

| Desirable AP*NET Characteristic | O*NET™ Characteristic |
| --- | --- |
| Uses a common language. | Uses a common language developed through extensive literature reviews and analyses. |
| Includes descriptors for a wide range of person attributes (e.g., interpersonal, problem solving). | Includes a comprehensive set of personal characteristic descriptors. |
| Includes descriptors for general work activities, skills, and knowledges that are relevant across jobs. | Includes cross-job descriptors that would need to be supplemented with Army-specific cross-job descriptors. |
| Includes occupation-specific descriptors (e.g.,specific tasks, equipment, and technology). | Includes a process for gathering occupation-specific descriptors.  Does not include task, equipment, or technology descriptors. |
| Includes descriptors at varying levels of specificity arranged hierarchically. | Includes hierarchically organized descriptors. |
| Includes a taxonomy of missions and linkages among missions, work activities, skills, and knowledge. | Does not include Army-specific variables. |
| Includes variables or aggregates of variables likely to be useful to commanders. | Does not include Army-specific variables. |
| Is linked to civilian occupational analysis databases. | Is linked to the Bureau of Labor Statistics databases. |
| Is automated and on-line. | Is planned to be automated and on-line. |
| Is coupled with a top-down future-oriented analysis procedure. | Does not include a built-in future-oriented job analysis approach. |

SOURCE:  Russell et al. (1995).

acteristics is shown in Table 6.11 (taken from the report). It can be seen from this table that some of the areas proposed for the AP*NET have corresponding characteristics in O*NET™, whereas others would require specific development for Army use. For example, O*NET™ uses a common language, includes a comprehensive set of personnel characteristic descriptors and cross-job descriptors, uses hierarchically organized descriptors, provides a process for gathering occupation specific descriptors, includes civilian jobs, and is planned for on-line automation. The Army would have to develop:

- Army-specific cross-job descriptors;
- Equipment and technology descriptors to be linked to tasks;
- A taxonomy of missions and linkages among missions, work activities, skills, and knowledge; this taxonomy will need to be included in the common language so that missions can be linked to tasks, skills, and knowledge;
- Variables of use to commanders; and
- Coupling with top-down, future-oriented job analysis procedures.

Russell et al. (1995) suggest that the Army build a prototype of AP*NET and then run a pilot test to identify development problems and to assess usefulness. In addition, their report makes the following long-term recommendations. First, develop procedures for assigning individuals to teams that optimize on multiple personnel considerations associated with readiness. Second, initiate studies to identify and measure individual and teamwork requirements for new missions. Third, develop performance measures to be used in career development, training, and job classification. Finally, develop simulation exercises to allow officers to run what-if scenarios based on various mixes of force capabilities. The key is to develop an occupational analysis system that efficiently links workforce capabilities with mission planning and provides the structure for recruiting, training, and assignment and promotion of personnel.

# 7

# Conclusions and Implications

As noted in Chapter 1, the nature of work and its role in society are changing in ways that have sparked considerable debate among social critics and scholars from multiple disciplines who study work. In this book we have analyzed how the nature of work and the occupational structure in the United States are changing and discussed the implications of these changes for how the tools of occupational analysis and classification can be used to bundle jobs together into work structures. We have attempted to sort out what is changing and what is not and to provide an interpretative framework that will aid organizational decision makers and members of the workforce as they go about making choices that will not only shape the world of work but also the future of numerous stakeholders. In this final chapter, we present our key findings, discuss their implications for decision makers in both the civilian and military sectors, and suggest directions for future research.

Our analysis has been guided by the following four broad themes: (1) there is increased diversity in the workforce and within occupations, (2) traditional occupational boundaries are becoming increasingly blurred, (3) the range of choices open to human resources managers and other decision makers about how to structure work appears to be increasing, and (4) there is a need for occupational analysts to think systematically about the range of forces influencing how work is structured. We start by summarizing the major findings underlying these themes, then we present implications for occupational analysis, for the military, and for policy and research.

Throughout this volume we have presented evidence that work is changing; that occupational analysis and classification systems such as the *Dictionary of Occupational Titles* are backward-looking and do not accurately represent the structure of work today; and that a better system is needed that adequately describes the current status of work, takes into account the forces that influence the nature of work, and can be easily and frequently updated as changes occur. A logical but more speculative extension of our analysis of the role of occupational analysis systems is the vision that, if these tools can be designed to be forward-looking, they can serve as analytic aids to decision makers in designing jobs and in creating human resource management policies. In the view of the committee, this is an area in which future research and experimentation should be vigorously pursued.

## SUMMARY OF THE EVIDENCE ON THE CHANGING NATURE OF WORK

### External Forces

#### Occupational Structures

Some of the changes in work documented in this volume are long-term and evolutionary in nature and some are of more recent origin. The most visible long-term changes can be seen in the evolution of the overall structure of occupations. The past several decades have witnessed a gradual expansion of technical and professional occupations, as blue-collar and farming occupations have declined. Managerial employment has remained relatively constant, although the work traditionally thought to be the province of managers is now being shared with employees across different levels of the occupational hierarchy, especially in cases in which high-performance work systems are being used. Both lower-level and professional service occupations are expanding.

These trends are likely to continue, although the pace of change will be gradual and evolutionary.

The cumulative effects of changes in the occupational structure should change the images of work that have been carried over from the days when industrial work was dominant. Yet old images, and the institutions and policies they fostered, may still too often dominate analysis and actions that affect work. Distinctions between blue-collar and managerial work and exempt and nonexempt employees continue to influence the characterizations of occupations and the policies and practices that govern work. These images and categories are no longer as useful as they once were for describing what workers actually do. Many rest on the cultural dichotomy between mental and manual work and on the notion that those who do the former rarely do the latter and vice versa. Although the distinction may have always been easier to make in theory than in practice, studies of work indicate that the distinction is less viable than ever. For instance, technicians' work is almost always marked by a fusion of sensorimotor skills and extensive knowledge of body of a scientific or technical knowledge. The diffusion of advanced information and control technologies and quality control techniques into factories has brought a significant amount of analytic and symbolic work onto the factory floor. As professionals and managers join the ranks of contingent employees, many shift from being paid salaries to being paid by the hour. Rethinking outmoded images and categories and their implications for how work is structured is critical to creating a more accurate and useful map for job designers, personnel managers, career counselors, and employers.

### Workforce Demographics

The demographic characteristics of the labor force continue to evolve and change. The dominant effect of these trends is to increase the diversity of the labor force, particularly with respect to age, education, race, sex, and marital status. For our purposes, the key feature of this increased diversity is that much of it is occurring within occupational groups and therefore may have significant implications for occupational analysis systems. Occupations are more likely than in the past to have a mix of workers of various tenures and workers earning different wages. This internal variation provides prima facie evidence that the correspondence between the jobs these workers are performing and the occupational classifications used to label them may be breaking down. Jobs are shaped by the interactive effect of (1) managerial and engineering choices with respect to work design and technology and (2) the knowledge, skills, abilities, and outlooks that individuals bring to the job. The increased variation in demographics that we are witnessing within any one occupation suggests the prospect of broader variation in job content. Going beyond this suggestive evidence, however, requires more intensive and direct analysis of the tasks workers are actually performing.

**Markets**

Changes in product and financial markets are producing changes in the nature of work. Product market forces are having two different effects. Increased domestic and global price competition has put pressure on labor costs. As a result, jobs that in the past paid high premiums but that can be supplied more cheaply in other countries or in domestic enterprises that pay competitive market rates are moving to these lower-cost environments and organizations. This is particularly true of semiskilled blue-collar work in both manufacturing and services. Wage competition is one cause of the restructuring experienced in American industry in recent years.

Along with increased price competition, markets have changed in ways that require increased capacity and speed in developing and introducing new and more varied products. This in turn increases pressures within organizations for flexibility in work organization and sets off an interrelated set of changes in organizational structures and human resource practices: specifically, flatter hierarchies, greater horizontal or cross-functional coordination through teams, and personnel policies designed to support increased flexibility and adaptability. The rapid change in product and service markets and product life cycles suggests that there is more rapid churning in the knowledge base required of employees who design and produce these goods and services.

Knowledge and skill obsolescence is likely to be of growing concern for occupational analysis.

Financial market pressures appear to have exerted an increased role in the structure of work in recent years. Increased shareholder activism has disciplined managers to be more efficient in the use of physical, financial, and human assets. Sometimes this has led to the shedding of operations and employees not central to the core competencies of the firm. Managerial compensation has been tied more tightly to firm performance, and downsizing (typically without success) has been used as a preemptive strategy to improve financial performance rather than as a last resort in response to financial crises (Cappelli, 1999; Cascio, 1993). As a result, employees across the occupational spectrum, from entry-level workers to top executives, have been exposed to risks of restructuring and downsizing. The increased volume and volatility of global capital also contributes to uncertainty. The effects of these financial forces have not, however, been well researched but represent another potentially important factor that should be brought into the analysis of how work is structured and changing in today's organizations. Deepening understanding of how financial markets and institutions affect the context and content of work and their consequences for firms, employees, and other stakeholders represents an important research priority for the future.

### Digitization: A New Technical Infrastructure

Changes in technology serve as another strong external influence on the structure of work and content of jobs. Innovations in digital technologies are clearly the most important technical developments of our time and are leading some to suggest that their ultimate effects will be equivalent to those of the first and second industrial revolutions. Although there is insufficient evidence to describe digitization as the kind of technological infrastructure that will produce a third industrial revolution, digital technologies are exerting profound effects on both the content of jobs and the occupational structure.

Like other major technical developments of the past, digital technologies are having at least three effects. First, they are changing the mix of jobs. New jobs such as computer scientists, programmers, and technicians have arisen while printing press operators, telephone and telegraphic operators, bank tellers, and other occupations that revolved around manual or mechanical means of processing information have declined. Second, by increasing the number of people across the occupational structure whose jobs involve the use of information processing technologies, the skills required to do many jobs have either increased or decreased. On balance, the evidence suggests that more workers are experiencing an increase than a decrease in skills as a result of digitization. Third, digitization is changing the types of skills needed, particularly by reducing some of the manual and sensory-based skills and increasing the analytic and information processing requirements of many production jobs and shifting the emphasis from mechanical manipulation to use of digital symbols and information.

If, as some believe, digitization represents a new technological infrastructure equivalent to the first and second industrial revolution, then the pace of technological change and its impacts on work and occupations may only be in its early stages and will probably accelerate in future years. Historians will have to make the judgment on this issue many years hence. In the meantime, to serve those who design occupational analysis systems and other decision makers, we require more direct observation and systematic sampling of the broad range of jobs affected by these technologies. We stress particularly the need to go beyond the debate over up- or down-skilling and to consider the extent of the reconfiguration of skills and the rapidity of change in the knowledge and skill content of jobs, both of which have important implications for occupational analysis and the design of education and training systems. Capturing the changes in the skill mix involves analysis of the changing analytical, interactive aspects of work as well as the degree of control or discretion associated with a change in technology.

## Organizational Restructuring

All of these external forces—changes in demographics, markets, and technologies—both partially cause and interact with the significant amount of organizational restructuring that has occurred in the past two decades. Restructuring has taken a variety of forms, but three are most salient in their effects on work structures and occupations: (1) downsizing, (2) flattening organizational hierarchies and greater reliance on teams for both vertical and horizontal coordination, and (3) changing employment practices and relations.

## Downsizing

Downsizing has led to increased job insecurity, particularly for managerial employees. When evaluating the evidence on changes in job security, it is important to differentiate changes in job stability from job displacement. Stability, as measured by average tenure, is a function of both voluntary and involuntary quits. The best evidence from national samples indicates that at best there has been only a rather small decline in overall job stability in the 1990s. Job displacement, (i.e., involuntary quits or terminations) however, shows a decline from a peak in the recession years of 1981-1983 but then an increase again in the early 1990s. Perceptions of job security, as reported in survey data from nationally representative samples in 1985 and 1995, also declined. Consistent with the evidence from the other surveys discussed above, the declines in job security have been greater for managerial employees than for others. For example, over two-thirds (68 percent) of employers reported in one survey that they provided employment security for managerial employees in the past but no longer do so.

Thus, changes in employment stability and security might be summarized as follows: (1) involuntary separations have increased, particularly for managerial employees in large firms, (2) voluntary quits have declined, and (3) perceptions of job security have declined. This does not, however, add up to a picture of the disappearance of all long-term jobs, as some popular reports on downsizing and restructuring suggest. It does, however, mean that the employment relationship has become more uncertain. At the same time that firms were downsizing, many were also hiring. This was particularly true of smaller firms. This high level of job churning increases the importance of having information about jobs across firm boundaries.

## Flattening Hierarchies

The flattening of hierarchies not only reduces the number of managerial employees but also delegates greater authority to lower-level employees. Middle managers have been among the hardest hit from both layoffs and the redefinition of work roles. Yet, surprisingly, few researchers have directly observed and reported on the actual nature of contemporary middle management work. (An exception is the work of Floyd and Wooldridge, 1996.) Instead, most of the studies we have used to assess the effects of restructuring on managerial work focus on the changes in employment conditions. These indeed are quite significant and visible. Job security—actual and perceived—has declined. Promotion opportunities appear to be limited and more open to competition from external candidates. Managerial compensation has been put at risk with higher percentages of pay coming in the form of performance bonuses. Use of outside contractors or consultants for specialized managerial tasks appears to have increased.

## Increased Use of Teams

Among the most visible changes in the structure of work is the increased use of teams. The use of teams reflects both a desire to move discretion to lower organizational levels and the need to enhance horizontal communications, coordination, and problem solving across traditional job and organizational boundaries. Because the use of teams has such a profound effect on how work is done, we discuss their role in greater detail below in the section devoted to changes in the content of work.

## Changing Employment Relations

Taken together, the changes in demographics, markets, technologies, and organizational structures and processes have exerted a profound effect on implicit and explicit employment contracts and conditions of employment in the civilian sector. Some of these effects were discussed earlier, and it is worth summarizing the key changes identified in previous chapters that bear on the nature of work and occupational structures. Wage inequality has increased and, for our purposes, it is important to note that a considerable portion of the increase is observed within occupations. Although job security and job stability have declined, the decline in job stability in recent years has been relatively slightly increased because of hiring by smaller firms and the reduction in voluntary quits. Union membership, particularly among blue-collar workers, continued its long-term decline simultaneously with significant innovation within union-management relations. Nonstandard employment, particularly with respect to the use of temporary workers and independent contractors, has increased.

## Changes in Content of Work

Whether the trends observed within organizations continue into the future, stabilize at their current levels, or reverse directions cannot be easily predicted since the direction and interaction of the forces acting on these trends (both exogenous and internal to the organization) are not easily predicted. However, taken together, the changes in the external and organizational contexts of work are having significant effects on the content of work. Indeed, the changes in work content summarized below are most profound in settings that are heavily exposed to or affected by the confluence of external and organizational contextual forces discussed above.

No single trend captures the changes in how work is done today. We characterize the dominant patterns observed by examining four aspects of work that are changing in significant ways in response to the changing contexts of work described above.

## Control and Discretion

The vertical division of labor is changing in organizations that have flattened their hierarchies, turned to team forms of work organization, and adopted human resource policies often described as "high-performance work systems." Moreover, in these settings blue-collar and managerial jobs are undergoing significant changes that are blurring the traditional lines of demarcation that separated these traditional categories. The autonomy of blue-collar workers has increased and some of the responsibilities traditionally reserved for supervisors, such as quality control, scheduling, and other operational responsibilities, have been delegated to nonsupervisory workers. This does not mean that blue-collar workers have gained significant control over what work is to be performed. Instead, their new authority involves greater discretion over how to do their work.

Although the trend in organizations that have adopted high-performance work systems is toward greater operational discretion, the trend is not universal. Discretion has been reduced in some service and manufacturing jobs that are designed to serve mass markets using more rational processes and information systems that management uses to monitor and control the pace and quality of work.

### Task Scope

Because of the foregoing changes associated with increasingly decentralized autonomy and control, the scope of blue-collar jobs is expanding, particularly in settings that make use of teams or other high-performance features to improve quality, innovation, and customer responsiveness. At the same time, the growth of specialized scientific knowledge has increased demand for specialists with state-of-the-art technical knowledge. Furthermore, professionals and technicians with specialized knowledge are more frequently working in interdependent, multidisciplinary, and cross-functional teams.

### Cognitive Complexity

The cognitive complexity of work appears to be increasing for blue-collar and service workers as a result of the technical and the organizational changes discussed above. The dominant trend is toward work that mixes physical and sensory skills with higher-level cognitive skills required by information processing technologies. Technical, professional, and managerial work has always entailed high cognitive content. Although there is no clear evidence to suggest that the cognitive levels of these jobs are changing in a significant way, the types of cognitive skills that are required may be changing, as we discuss below.

## Interactive and Relationship Requirements

Interpersonal interactions are becoming more important in many jobs. Interdependence and more direct interactions with other employees, customers, clients, and patients increase the importance of both substantive and emotional interactions that require skills in communications, problem solving, and negotiations.

## THE BROADER DEBATES ABOUT JOBS AND WORK

One purpose of this study was to clarify, with the best evidence available, some of the popular debates about the future of work and provide a richer interpretive framework for tracking changes in work in the future. We summarize our judgments on these issues here.

### Singular Trends Versus Constrained Choices

The evidence cautions us against making definitive statements about unidirectional trends. Although some trends do appear to be dominant, too much emphasis has heretofore been placed on identifying single trajectories and grand predictions regarding the future of work. Instead, we see increased variance within occupations and multiple options for shaping jobs and for grouping them into occupations. Thus the future of jobs will not be determined solely by the forces of technology, demographics, or markets but by the interaction of these forces with the strategies, missions, organizational structures, and employment policies that decision makers implement in specific settings. Choice remains important even when options are constrained by external events and when consequences for organizations, individuals, and society are imperfectly predictable.

For example, especially in settings in which markets are uncertain and goals are unclear, it is critical for people to coordinate with each other and communicate with clients, even though it might be possible to use information technology to support and rationalize work. Work in such settings is likely to be more productive if it allows high discretion, flexibility, and the opportunity to work in teams to solve problems, analyze data, and negotiate over courses of action or the meaning of information. In such environments, failing to structure work in this way will likely result in lower levels of workers' performance and a less rewarding experience for the individuals involved. Changing some but not all of these features may produce mixed results and perhaps unstable arrangements.

The available evidence on the effects of high-performance work systems illustrates this point. Numerous quantitative studies from different industries suggest that when work practices and human resource policies are combined in ways that complement each other, the bundling of tasks has significant, positive effects on productivity, quality, and profitability. Although the evidence is less substantial, proper bundling also appears to result in higher levels of job satisfaction, more employee learning, and higher wages for blue-collar workers. But since these high-performance work systems shift responsibilities formerly held by supervisors and middle managers to teams of lower-level workers, they are likely to reduce the number of middle management jobs and change the nature of supervisory duties. This, in turn, renders obsolete public policy doctrines that depend on a clear line of demarcation between the duties of managers and workers. The adoption of these systems is most likely in firms that are under competitive pressures to produce high productivity and quality.

Thus, although firms do exercise discretion over the decision to implement a high-performance work system, their choice has predictable (although not certain) consequences for at least three stakeholders: the firm and its shareholders, the employees who are affected by the shift to these systems, and public policy makers who are responsible for laws and regulations governing employment relations. Only by considering the full range of factors in the framework presented early in this volume (Figure 1.1) can the consequences of the relationship between driving forces and choices about work content be understood. We caution, however, that research that examines a full range of forces, choices, and consequences is still in an early developmental stage. Most of the evidence to date comes from a rather narrow range of industries and lower-level occupations. Strengthening the understanding of these issues and the confidence of the direction and magnitudes of these effects across a broader range of occupations and industries is an important priority for future research.

### The End of Jobs?

Nothing in the data we have examined would support the conclusion that all the changes in today's workplace add up to "the end of jobs" in any sense of this term. The conditions and content of work are certainly changing in sometimes dramatic ways, but the vast majority of people in America who want or need to work remain employed. Employment, labor force participation rates, and hours of work have either increased or remained stable in recent years. There is no compelling evidence to suggest that this will change in the future. Moreover, the history of technology repeatedly shows that, even when large numbers of individual workers are driven from particular jobs as a result of a shift in the demand for labor, aggregate demand for workers does not decline because of technical change.

Although claims about the end of work can be dismissed as hyperbole, the various voices debating this issue may signal a deeper phenomenon. Their concerns may be a symptom of the general perception that we are in the midst of a third industrial revolution driven by a change in technical infrastructure associated with digitization. Historians will eventually decide whether or not the changes in work opened up by digital technologies warrant the label of an industrial revolution, but the evidence that we have summarized is sufficient to suggest that microelectronic technologies are having profound effects on work and occupational structures. Their full effects are felt when they are combined with changes in organizational strategies and employment practices. The potential magnitude and importance of these changes make it essential that occupational analysis systems be both comprehensive and flexible enough to track these changes and provide decision makers with the information they need to mesh technical changes with work structures that meet the needs of the various stakeholders involved.

## A Transformation of Work?

Taken alone, none of the changes or trends we have discussed in this book constitutes anything that could be characterized as a transformation of work. But when combined, as we have seen in some settings, they may lead to new conditions and possibilities that some might characterize as a transformation. Indeed, one of our objectives has been to develop a framework that researchers, organizational decision makers, advisers, occupational analysts, and individuals can use as they experiment with designing work, occupational structures, and employment policies. The absence of a clearly articulated framework that includes the full range of forces shown in Figure 1.1 has limited our ability to assess the combined effects of various changes on individuals, organizations, and society itself. We believe that the social and organizational implications of the combination of changes that we identified need to be examined more fully and systematically by decision makers in both the civilian and military sectors. By explicitly taking into account the full range of factors that shape how work is done, we believe that decision makers have the opportunity to develop more effective alternative work structures that could potentially meet a broad range of needs and interests.

## IMPLICATIONS FOR SYSTEMS OF OCCUPATIONAL ANALYSIS AND CLASSIFICATION

To adequately track the changing nature of work, occupational analysis and classification systems must take into account the *attributes* of the persons who perform work, the *processes* by which they perform it, the *outputs* they produce within the dynamic economic, demographic, technological, and the organizational *contexts* that affect all three. To achieve this objective, occupational analysis and classification systems must widen their traditional scope of attention as well as deepen their level of descriptive detail to capture both the range of relevant factors and the distinctions between jobs and occupations that might otherwise go unnoticed. They must also be sensitive to the greater variance in how work is done within occupations today. Rather than provide a single description of a given job, an adequate system for occupational analysis may need to attend to various alternatives for structuring work in a given job family as well as to the attributes and skill requirements associated with these alternatives. Shifting from a backward-looking to a forward-looking system that will aid decision makers in designing work structures will also require occupational analysts to consider the human resource and organizational practices needed to support alternative ways of structuring work. By being flexible enough to address future changes in the context and content of work, occupational analysis and classification can contribute not only to the description of work, but also to research that interprets and predicts changes, and to the work designs that anticipate those changes.

As work becomes increasingly team-based, it changes the mix of skills possessed by a typical worker and blurs traditional demarcations across occupations. Changing economic conditions and conceptions of the employment are also leading many workers into nontraditional employment arrangements such as entrepreneurial, part-time, contingent, or contract work and self-employment. These developments have implications for methods and techniques of occupational analysis, because they call into question what is meant by a "job."

To account for people factors, job analysis inventories must become sufficiently detailed to describe such work attributes as abstract analytical work, skill in the use of information technology, teamwork competencies, and skill in performing emotional work. The job analysis process should include effective definitions and tools to identify skills required to perform information technology intensive work, interactive and emotional work, and team tasks and interactions. Process refinements are also needed with respect to whom to sample when analyzing a job or occupation, how to reach them, which questions to ask, what survey technology to use, and how often to update the analysis.

To clearly describe work processes and products, definitional refinements are required for such questions as: What is a "job"? What does a job title indicate about job content? How can title-based task inventories be created when incumbents share tasks across titles? Job activities must be precisely defined. Occupational analysis and classification process improvements relevant to analysis of work processes and outputs include deciding which observers and analysts will most effectively complete which inventory, and how often the job analysis information should be updated, especially with respect to technology used on the job (Cascio, 1995).

In summary, a more contingent approach to describing jobs is needed that can take account of different systems for organizing work and of work systems that are undergoing continuous improvement or more rapid change. An example might help illustrate what we have in mind. Consider the job of an automobile assembly line worker. If organized in the traditional way, an assembler will be assigned a narrow, highly specialized task with specific rules governing how the job is to be performed. If organized in a team-based or high-performance work system, the same job is likely to encompass a wide range of physical, analytic, and interactive tasks that include considerable individual and collective discretion over how the work is to be done. The same would be true of customer service workers. As reported in Chapter 4, the scope, discretion, cognitive complexity, and interactive content of these jobs vary considerably depending on the type of customer the worker serves. And as product and service offerings continuously change with new technologies and customer demands, by implication so too do the information processing requirements of the customer service jobs. One implication of this is that much more research is needed to identify the critical contingencies (e.g., customers, product life cycles, production organization) and the driving forces that shape the choice of alternative work systems.

Analysis of changing work contexts is likely to require inclusion of the home environments or other alternative work environments of employees (e.g., telecommuters) when gathering context information. Employment relationships and organizational structures should be carefully defined and examined to capture changing understandings of who are considered employees (e.g., distinctions between contract and regular employees). Improvements in the analytical process for examining context factors should involve means to effectively include incumbents with alternative work schedules in the job analysis sampling plan.

General process challenges include the adaptation and application of technology to enable effective and efficient gathering, updating, and analysis of job information on a large scale. Finally, research suggests that workers, particularly highly skilled service workers, often have considerable leeway in defining the scope of services provided and the manner of service delivery. Such individual work discretion presents a challenge to task-oriented occupational analysis methods designed according to the premise that work processes and products are mostly prescribed by job titles. Occupational analysis and classification should, therefore, be sufficiently flexible to address unpredictable combinations of job attributes.

In the committee's view, O\*NET™ offers several important advances over prior systems in its organization of job description variables and associated data collection instruments, in its electronic databases with job incumbent and occupational analyst ratings, and in the initial technical evaluations. If fully developed and widely used by practitioners who add their own features to the system, we anticipate that it can serve the functions called for here.

- First, O\*NET™ is the first available system with planned national scope that brings together the most current category and enumerative systems and the most comprehensive descriptive analytical systems and makes the data readily accessible in electronic format.
- Second, O\*NET™ has a theoretically informed and initially validated content model with a more detailed set of job descriptors than other available systems.
- Third, the O\*NET™ database can be accessed and used through multiple windows or modes, including entering with job titles or occupations at varying levels of hierarchical detail, but also entering at the level of work descriptors (i.e., knowledge, skills, abilities, other contextual factors). The latter window of access is extremely important in a world of work that is changing. It allows the analyst or user to build up inductively to the level of job or occupation, in contrast to systems that proceed deductively,

starting with a job or occupational category that is anchored in the past and may not be current in its ratings or job descriptive information. O*NET™ could be developed into a decision support tool that allows analysts to compare different models for organizing work, to generate a list of complementary changes needed to support these models, and to project the consequences of these alternatives for the outcomes of central interest to different stakeholders. This feature is perhaps one of the major developments of the O*NET™ prototype.

• Fourth, O*NET™ offers a significant improvement over earlier systems, particularly ones based on the *Dictionary of Occupational Titles*, in the ease of conducting cross-occupational analyses and comparisons.

• Fifth, by utilizing the cross-walks supplied by the National Occupational Information Coordinating Committee, the O*NET™ system allows mapping to other major category and enumerative systems, including military occupational specialties and the Standard Occupational System.

Based on these advances, the committee recommends that O*NET™ should continue to be developed as a fully operational system for use in both civilian and military sectors.

## IMPLICATIONS FOR THE ARMY

Since most of the evidence we have discussed pertains to the civilian sector, it would be inappropriate to generalize our findings or conclusions directly to the Army or other branches of the military. Rather than draw such generalizations, we sought to organize our review of the changes in civilian occupations and organizations in a way that would provide the Army with a framework for examining its own work structures and occupational analysis systems. As the material in Chapter 6 shows, the Army is experiencing a number of changes in the context of work that parallel changes experienced in the private economy. We suspect that these developments will also create pressures for change in the structure and content of soldiers' work. These pressures should create opportunities for the military commanders to adjust existing work systems, should they choose to take advantage of them.

The Army's work structure is the basis for selecting, training, organizing, and managing personnel to meet mission requirements. The result of changing mission requirements has been the development of a smaller, more flexible force with a wider range of fighting skills—as well as new skills in negotiation and interpersonal interaction. The increased diversity of Army missions coupled with downsizing has led to the creation of teams composed of individuals from different work cultures with different skills.

## Diversity

The growing role of women in the military, the increasing age and family responsibilities of military personnel, the growing use of units composed of regulars, reservists, and civilians, and the increasingly prevalence of joint missions with military units from other countries indicate that diversity will be a salient feature of military work. It is beyond the scope of our work to judge whether or not officers and soldiers are being trained adequately to lead and function effectively in these diverse settings. Given these developments, however, we believe that a thorough analysis of this question should be undertaken.

## Increased Use of Advanced Technology

The Army views technology as a key strategic and tactical resource. Yet there is still a tendency to view the choice, design, and use of technologies as somewhat separate. There is a need to better integrate the design of technology and the design of work structures. Those who design and purchase new technical systems need to be informed by and work in tandem with those who design work structures. The key lesson from the civilian sector is that the maximum benefits of a technology come from its effective integration with work systems and human resource practices.

Another implication of the Army's use of advanced technologies is the probable upskilling of jobs. It is likely that the qualifications required of personnel in the leaner military of the future will parallel the sophistication of the systems that are fielded. Occupational analysis and classification systems must capture these qualifications.

## Changing Missions

Rapid expansion (on-demand increases in the size of force and its range of capabilities) and tailoring (matching force capabilities to mission needs) will be required to respond to sudden events that do not resemble traditional military actions. Thus, the work structure itself is likely to become an increasingly important weapon or tool—a part of the information technology that must be used to speedily identify, locate, and assign large numbers of personnel for specific missions.

The trend toward joint operations involving hybrid units may require the branches of the military to develop common work structures or at least structures that can be more easily meshed. New missions—particularly those that involve extensive interaction with civilians—will require new tasks, knowledge, skills and abilities, and forms of organizing. For example, social skills will prove increasingly important for dealing with civilians and their leaders when Army personnel must adapt to peacekeeping, community service, and similar roles. These tasks also may require fluency in the native language; cross-cultural sensitivity and knowledge; interactive and emotional skills; and the ability to engage enemies who do not follow traditional rules of war. Work structures must include descriptors for these tasks and the requisite knowledge, skills, and abilities.

Extremely high tempo and a strategy that allows attack from multiple positions simultaneously will require commanders who possess suitable decision-making, teamwork, and problem-solving skills; work structures must therefore include descriptors for these skills. Commanders will be supported by different organizational structures that delegate responsibilities and authority to lower ranks, and by technologies that permit such delegation by rapid collection, transmission, and analysis of information. Commanders must be able to plan, decide, and execute commands extremely rapidly and with great flexibility— rapidly grasping changes in situations and exercising judgment while reasoning under uncertainty. Moreover, as we will note below, the need for these skills is not limited to officers. The same forces that drive delegation of greater decision-making authority and a blurring of the manager-worker boundary in the civilian sector are at work on the officer-soldier boundary in the work of the Army.

## Downsizing

The downsizing of the military forces also creates pressures to delegate to lower-level soldiers tasks and decisions traditionally embedded in officer ranks. This pressure may be heightened by technology (e.g., distributed battlefield technologies) that encourages such delegation of authority and responsibility. Layers of supervision may be reduced. The officer-soldier distinction may experience some blurring similar to that affecting the blue-collar-managerial distinction in the civilian sector. Attention should be given to the implications of these new roles for personnel management. There are at least two possible cases in which previously distinct jobs may be at least partially melded in the future. One distinction is that between operational users (e.g., tank crews, helicopter crews) and maintainers. Another distinction is that between commanders and subordinates.

Although the military's new strategic emphasis on speed and flexibility of response has implications for a wider distribution of information and work responsibilities, there is little serious consideration being given to flatter, nonhierarchical organizational structures. Nonetheless, with fewer personnel, there may be a need for individual soldiers to have a broader range of skills than in the past or to work with a more diverse group of individuals from various sectors of the Department of Defense. The work structure should be flexible enough to adapt to these changes, and it must facilitate a rapid response to a wide variety of situations and team configurations.

## Systemic Effects

One of the most important implications suggested by the framework used to organize our analysis and the evidence presented about the factors that are interrelated to work structures is that Army decision makers need to see the design of jobs, work structures, and occupations as tightly linked to their changing missions, technologies, workforce demographics and family structures, and employment practices. The committee therefore recommends that Army decision makers think about the interconnections among these factors and take them into account in structuring work to meet the mission requirements and the needs of those who will be part of the Army of the future.

## Application of O*NET™

The Army's ability to efficiently manage its personnel, in the complex and rapidly changing contexts expected for future missions, would be enhanced by an occupational analysis system that efficiently links workforce capabilities with mission planning and provides the structure for recruiting, training, assignment, and promotion of personnel. Such a system would provide all the information needed for such tasks as assembling a special operation in the field or for developing training requirements for a combined military occupational specialty.

Having considered the advantages of O*NET™, the committee sees that it offers promise for meeting the future occupational analysis needs of the Army. AP*NET, an adaptation of O*NET™ proposed by Russell et al. (1995), has several useful features. It will (1) be useful to manpower, personnel, and training professionals and to Army commanders; (2) contain linked readiness, occupations, and training databases that allow easy access to descriptions of training courses that teach a particular skill, to lists of soldiers who have skills and abilities relevant to a particular type of mission, and to Army jobs that have similar requirements; (3) have a menu-driven, user-oriented interface that allows users to access data at the level of aggregation and specificity that is best suited to the application. In adapting O*NET™, the Army would have to develop Army-specific cross-job descriptors; equipment and technology descriptors to be linked to tasks; a taxonomy of missions and linkages among missions, work activities, skills, and knowledge; variables of use to commanders; and coupling with top-down, future-oriented job analysis procedures.

The committee recommends that the Army consider building a prototype of a system whose functional capabilities include those in the AP*NET concept.

## IMPLICATIONS FOR RESEARCH

### Need for Multidisciplinary Studies of Work

Our committee used a blend of theoretical perspectives and methodologies and data drawn from multiple disciplines to assess how work is changing and its implications. In doing so, we both broadened the perspectives most scholars and practitioners use to study work and to make decisions that affect work structures and their consequences. In our view, this type of cross-disciplinary collaboration is essential to future progress in this field. However, we do not suggest that all individual studies should abandon their disciplinary focus or traditions. Instead, communication and dialogue across disciplines is needed to inform both the framing of questions and the interpretation of results from multiple disciplines.

### Rethinking Images of Work and Occupations

We have been critical of both the popular and analytic images and categories typically used to characterize and study work and occupations that have been carried over from an earlier era when industrial and manufacturing work dominated the economy. These images and categories reflect the historical periods in which they were formed. To gain the full advantage of the opportunities available from new technologies and organizational forms and the changes in the characteristics of the labor force, the images need updating to better reflect: (1) the diversity of the workforce, (2) the dominance of the service economy, (3) the growing role of cognition and analysis, interactions and relationships, and digital technologies in the work people do, and (4) the blurring of the traditional boundaries across which work was divided in the industrial era. The blue-collar-managerial divide in particular no longer captures what people do at work. How to adapt practices, institutions, and public policies that rely on this divide or the other outmoded images are major issues for future study and action.

## Need to Study What Workers Do

Throughout this volume we have stressed the need for more intensive direct observation of what workers actually do in their jobs today. Changing the images of work and going beyond abstract arguments about trends in skills requires detailed and rich description and data reported from direct experiences of workers. Thus the sociological and anthropological traditions of observing and participating in real work settings and producing detailed narratives describing the actual experiences of workers need to be encouraged, with the objective of updating perspectives on work. But to be representative, these studies must examine the full array of occupations and workers found in the labor force today. Researchers are especially limited in their ability to describe what managers do at work today because it is difficult to measure. Furthermore, sociologists, industrial relations experts, anthropologists, and others continue to focus their efforts on more easily quantifiable jobs in lower-level occupational groups. It is also important to examine ways of integrating data describing what workers do from other disciplines, such as industrial and organizational psychology and human factors.

## Need for a National Database on Work

Direct observation and in-depth description of what workers do is a necessary but not sufficient input to update and continue to monitor changes in the aggregate structures of work and the content of jobs. To do this requires a national sample representative of the labor force. This type of data collection is needed both to complete the data collection and analysis needed to make O*NET™ operational and to realize its potential and to track systematically the changes in work and their consequences for organizations, individuals, and society.

## Need to Study Occupational Analysis Tools as Aids to Decision Makers

As noted earlier, the committee's vision is of a forward-looking occupational analysis system that can be used by decision makers to monitor changes in work, design new jobs, formulate effective personnel policies, and provide timely career counseling. Advances in technology that allow for the consideration of large numbers of variables in a relational database have made it possible to include information not only about jobs and skills, knowledge and abilities, but also about the organizational and environmental forces that influence work. Furthermore, it is now possible to display and combine data to develop what-if scenarios as an aid to job design. In the committee's view, the use of occupational analysis tools to shape work is an extremely important and fruitful area for research and experimentation.

## IMPLICATIONS FOR POLICY

Throughout this study, we noted that the laws and institutions governing work and employment largely reflect their industrial-era origins. It goes well beyond the scope of this effort to suggest what changes are needed to update employment laws and institutions to better support work and employment relations today. As we note in the introduction, there are good reasons to believe that the current structures and content of workplace regulations may have two adverse effects. First they may force the organization of work into outmoded categories. Second, because of increased complexity, they may produce frustrations on the part of workers and employees. Thus, although they may be beyond the scope of this analysis, these issues warrant study of their own at some point in the near future. Furthermore, this book may provide a starting point for the analysis of the role of law by presenting data on how work has changed since the basic legal framework governing employment relations was enacted.

# References

Adler, P.S.
  1992   *Technology and the Future of Work.* New York: Oxford University Press.
  1993   The "learning bureaucracy": New United Motor Manufacturing, Inc. Pp. 111-194 in *Research in Organizational Behavior*, Vol. 15, B.M. Staw and L.L. Cummings, eds. Greenwich, CT: JAI Press.
Advisory Panel for the Dictionary of Occupational Titles (APDOT)
  1993   *The New DOT: A Database of Occupational Titles for the Twenty-First Century* (Final Report). U.S. Department of Labor Employment and Training Administration. Washington, DC: U.S. Employment Service.
Albrecht, K., and L.J. Bradford
  1989   *The Service Advantage: How to Identify and Fulfill Customer Needs.* Homewood, IL: Dow Jones-Irwin.
Allen, T.J., and R. Katz
  1986   The dual ladder: Motivational solution or managerial delusion. *R&D Management* 16:185-197.
American Institutes for Research
  1997   *Occupational Information Network (O*NET) Technical Executive Summary.* Salt Lake City, UT: Department of Employment Security.
American Management Association
  1994   *Survey on Basic Skills Testing and Training.* New York: American Management Association.
  1996   *1996 AMA Survey on Downsizing, Job Elimination, and Job Creation.* New York: American Management Association.
Ames, C., and J. Hlavacek
  1989   *Market Driven Management: Prescriptions for Survival in a Turbulent World.* Homewood, IL: Dow Jones-Irwin.

Anderson, G.
  1988   *The White Blouse Revolution: Female Office Workers Since 1870.* Manchester and New York: Manchester University Press.
Anderson, J.R.
  1993   Problem solving and learning. *American Psychologist* 48:35-44.
Aoki, M.
  1988   *Information, Incentives, and Bargaining in the Japanese Economy.* New York: Cambridge University Press.
Appelbaum, E., and R. Batt
  1994   *Transforming Work Systems in the United States.* Ithaca, NY: ILR Press.
Appelbaum, E., and P. Berg
  1999   Hierarchical organization and horizontal coordination: Evidence from a worker. In *The New Relationship: Human Capital in the American Corporation*, Margaret Blair and Thomas Kochan, eds. Washington, DC: Brookings.
Applebaum, H.
  1992   *The Concept of Work: Ancient, Medieval, and Modern.* Albany, NY: SUNY Press.
Applebaum, E., T. Bailey, P. Berg, and A.L. Kalleberg
  Forthcoming   *High Performance Work Systems: Effects on Manufacturing Plants and Workers.* Ithaca, NY: Cornell University Press.
Aronowitz, S., and J. Cutler
  1998   *Post-Work: The Wages of Cybernation.* New York: Routledge.
Arthur, M.B., and D.M. Rousseau
  1996   *The Boundaryless Career: A New Employment Principle for a New Organizational Era.* New York: Oxford University Press.
Ash, R.A., and S.L. Edgell
  1975   A note on the readability of the Position Analysis Questionnaire (PAQ). *Journal of Applied Psychology* 60:765-766.
Association of the United States Army
  1997   *Profile of the Army: A Reference Handbook.* Institute of Land Warfare. Arlington, VA: Association of the United States Army.
Attewell, P.
  1987   The de-skilling controversy. *Work and Occupations* 14:323-46.
  1992   Skill and occupational changes in U.S. manufacturing. Pp. 46-88 in *Technology and the Future of Work*, Paul S. Adler, ed. New York: Oxford University Press.
Baba, M.L.
  1995   Work and technology in modern industry: The creative frontier. Pp. 120-146 in *Meanings of Work: Considerations for the Twenty-First Century*, F.C. Gamst, ed. Albany, NY: SUNY Press.
Babson, S.
  1995   *Lean Work: Empowerment and Exploitation in the Global Auto Industry.* Detroit, MI: Wayne State University.

Bachu, A.
  1995   *Fertility of American Women: June 1994.* Current Population Reports, P20-482. Washington, DC: U.S. Government Printing Office.
Bailey, T., and A. Bernhardt
  1997   In search of the high road in a low-wage industry. *Politics and Society* 25(2):179-201.
Bailyn, L.
  1985   Autonomy in the industrial R&D lab. *Human Resource Management* 24:129-46.
  1993   *Breaking the Mold: Women, Men, and Time in the Corporate World.* New York The Free Press.
Bailyn, L., and J.T. Lynch
  1983   Engineering as a life-long career: Its meaning, its satisfactions, its difficulties. *Journal of Occupational Behavior* 4:263-83.
Baker, G., M. Gibbs, and B. Holmstron
  1994   The internal economics of the firm, evidence from personnel data. *Quarterly Journal of Economics* 109(4)November:881-920.
Bakker, B.F.M.
  1993   The Netherlands standard classification of occupations 1992. Pp. 272-281 in *Proceedings of the International Occupational Classification Conference.* Bureau of Labor Statistics Report 833. Washington, DC: U.S. Department of Labor.
Ballentine, R.D., J.W. Cunningham, and W.E. Wimpee
  1992   Air Force enlisted job clusters: An exploration in numerical job classification. *Military Psychology* 4:87-102.
Baran, B.
  1987   The technological transformation of white-collar work: A case study of the insurance industry. Pp. 25-62 in *Computer Chips and Paper Clips: Technology and Women's Employment,* H.I. Hartman, ed. Committee on Women's Employment, National Research Council. Washington, DC: National Academy Press.
Barker, J.R.
  1993   Tightening the iron cage: Concertive control in self-managing teams. *Administrative Science Quarterly* 38:408-37.
Barley, S.R.
  1988   Technology, power and the social organization of work: Toward a pragmatic theory of skilling and deskilling. Pp. 33-80 in *Research in the Sociology of Organizations,* Volume 6, N. DiTomaso and S.B. Bacharach, eds. Greenwich, CT: JAI Press.
  1990   The alignment of technology and structure through roles and networks. *Administrative Science Quarterly* 35:61-103.
  1996a  *The New World of Work.* London: British North-American Research Committee.
  1996b  Technicians in the workplace: Ethnographic evidence for bringing work into organization studies. *Administrative Science Quarterly* 41:404-41.

Bassi, L.J., A.L.Gallagher, and E. Schroer
1996   *The ASTD Training Data Book*. Alexandria, VA:  American Society for Training and Development.
Bassi, L.J., G.S. Benson, M.E. Van Buren, and R. Bugarin
1997   *1997 Human Performance Practices Report*. Alexandria, VA:  American Society for Training and Development.
Batt, R.
1996   From bureaucracy to enterprise? The changing jobs and careers of managers in telecommunications services.  Pp. 55-80 in *Broken Ladders: Managerial Careers in Transition*, P. Osterman, ed.  New York:  Oxford University Press.
1999a  Work organization, technology, and performance in customer services and sales. *Industrial and Labor Relations Review*.  July.
1999b  Strategic Segmentation: Linking Business and Human Resource Strategy in Customer Services.  Working Paper.  Center for Advanced Human Resource Studies, ILR School, Cornell University.
Batt, R., and J. Keefe
1999   Human resource and employment practices in telecommunications services: 1980-1998. In *Employment Strategies: Why Similar Employers Manage Differently*. Oxford, England:  Oxford University Press.
Beckman, S.
1996   Evolution of management roles in a networked organization:  An insider's view of the Hewlett-Packard Corporation.  Pp. 155-184 in *Broken Ladders: Managerial Careers in the New Economy*, P. Osterman, ed.  New York:  Oxford University Press.
Bell, D.
1973   *The Coming of Post-Industrial Society: A Venture in Social Forecasting*.  New York:  Basic Books.
Beller, A.
1984   Trends in occupational segregation by sex and race, 1960-1981.  In *Sex Segregation in the Workplace, Trends, Explanations, Remedies*, B.F. Reskin, ed.  Committee on Women's Employment, National Research Council.  Washington, DC:  National Academy Press.
Belzer, M.
1994   The motor carrier industry:  Truckers and teamsters under siege.  In *Contemporary Collective Bargaining in the Private Sector*, P. Voos, ed.  Madison, WI:  Industrial Relations Research Association.
Bennet, W., Jr., H.W. Ruck, and R.C. Page, eds.
1996   Special issue: Military occupational analysis. *Military Psychology* 8:115-265.
Berggren, C.
1992   *Alternative to Lean Production:  Work Organization in the Swedish Auto Industry*. Ithaca, NY:  Cornell University Press.
Biggart, N.
1988   *Charismatic Capitalism: Direct Selling Organizations in America*.  Chicago:  University of Chicago Press.

Binkin, M.
1994 Military technology and Army manpower: Do smart weapons require smart soldiers? Pp. 169-187 in *Marching Toward the 21st Century: Military Manpower and Recruiting*, M.J. Eitelberg and S.L. Mehay, eds. Westport, CT: Greenwood Press.

Binkin, M., and M.J. Eitelberg
1982 *Blacks and the Military*. Washington, DC: Brookings Institution.
1986 Women and minorities in the all-volunteer force. In *The All-Volunteer Force After a Decade*, W. Bowman, R. Little, and G.T. Sicilia, eds. Elmsford, NY: Pergamon-Brassey's.

Black, S.A., and L.M. Lynch
1997 How to Compete: The Impact of Workplace Practices and Information on Productivity. Working Paper. Cambridge, MA: National Bureau of Economic Research.

Blackburn, M.L., D.E. Bloom, and R.B. Freeman
1990 The declining economic position of less skilled American men. Pp. 31-76 in *A Future of Lousy Jobs? The Changing Structure of U.S. Wages*, G. Burtless, ed. Washington, DC: The Brookings Institution.

Blauner, R.
1964 *Alienation and Freedom: The Factory Worker and His Industry*. Chicago: University of Chicago Press.

Block, F.
1990 *Postindustrial Possibilities: A Critique of Economic Discourse*. Berkeley: University of California Press.

Bluestone, B., P. Hanna, S. Kuhn, and L. Moore
1981 *The Retail Revolution: Market Transformation, Investment, and Labor in the Modern Department Store*. Boston: Auburn House Publishing Co.

Boisjoly, J., G.J. Duncan, and T. Smeeding
1994 Have Highly-Skilled Workers Fallen from Grace? The Shifting Burdens of Involuntary Job Losses from 1968 to 1992. Unpublished manuscript. University of Quebec (Rimouski).

Booher, H.R., ed.
1990 *MANPRINT: An Approach to Systems Integration*. New York: Van Nostrand Reinhold.

Bookbinder, S.M.
1996 The employee perspective. In *The New Deal in Employment Relationships: A Council Report*. Report No. 1162-96-CR. New York: The Conference Board.

Boris, E.
1994 *Home to Work: Motherhood and the Politics of Industrial Homework in the United States*. New York: Cambridge University Press.

Bound, J., and G. Johnson
1992 Changes in the structure of wages in the 1980's: An evaluation of alternative explanations. *American Economic Review* 82(3):371-392.

Bowen, D.E., and B. Schneider
  1988   Services marketing and management: Implications for organizational
         behavior. Pp. 43-80 in *Research in Organizational Behavior 10*, B.M. Staw
         and L.L. Cummings, eds. Greenwich, CT: JAI Press.
Braverman, H.
  1973   *Labor and Monopoly Capital.* New York: Monthly Labor Review Press.
  1974   *Labor and Monopoly Capital: The Degradation of Work in the Twentieth Cen-
         tury.* New York: Monthly Review Press.
Bridges, W.
  1994   *Job Shift: How to Prosper in a Workpalce Without Jobs.* Reading, MA:
         Addision Wesley.
Brynjolfsson, E., and S. Yang
  1996   Information technology and productivity: A review of the literature.
         *Advances in Computers* 43:179-214.
Bureau of Labor Statistics.
  1996   BLS reports on the amount of employer-provided training. July 19 press
         release. Bureau of Labor Statistics, Washington, DC.
Burtless, G.
  1995   International trade and the rise of earnings inequality. *Journal of Eco-
         nomic Literature* 33:800-816.
Butterfield, S.
  1985   *Amway: The Cult of Free Enterprise.* Boston: South End Press.
Callaghan, P., and H. Hartmann
  1991   *Contingent Work: A Chart Book on Part-Time and Temporary Employment.*
         Washington, DC: Economic Policy Institute.
Cannon-Bowers, J.A., S.I. Tannenbaum, E. Salas, and C.E. Volpe
  1995   Defining competencies and establishing team training requirements.
         Pp. 333-380 in *Team Effectiveness and Decision Making in Organizations*, R.
         Guzzo, E. Salas, and Associates, eds. San Francisco: Jossey-Bass.
Cappelli, P.
  1999   *The New Deal at Work.* Cambridge, MA: Harvard Business School Press.
Cappelli, P., L. Bassi, H. Katz, D. Knoke, P. Osterman, and M. Useem
  1997   *Change at Work.* New York: Oxford University Press.
Carey, M., and A. Eck
  1984   How workers get their training. *Occupational Outlook Quarterly* Win-
         ter:3-21.
Carre, F.
  1997   Workplace Reorganization in the Insurance Industry. Technical Paper.
         Economic Policy Institute. August.
Carre, F., and P. Joshi
  1997   Building Stability for Transient Workforces: Exploring the Possibilities
         of Intermediary Institutions Helping Workers Cope with Market Insta-
         bility. Working Paper No. 1. Cambridge, MA: Radcliffe College Public
         Policy Institute.

Cascio, W.F.
   1993   Downsizing: What do we know? What have we learned? *Academy of Management Executives* 7(1):95-104.
   1995   Whither industrial and organizational psychology in a changing world of work? *American Psychologist* 50:928-939.
Cascio, W.F., C. Young, and J.R. Morris
   1997   Financial consequences of employment-change decisions in major U.S. corporations. *Academy of Management Journal* 40(5):1175-1189.
Caudron, S.
   1993   Are self-directed teams right for your company? *Personnel Journal* 72(12):76-84.
Chandler, A.D., Jr.
   1977   *The Visible Hand: The Managerial Revolution in American Business.* Boston: Harvard University Press.
Chatman, J.A., J.T. Polzer, S.G. Barsade, and M.A. Neale
   1998   Being different yet feeling similar: The influence of demographic composition and organizational culture on work processes and outcomes. *Administrative Science Quarterly* 43:749-780.
Chi, M.H., R. Glaser, and M.J. Farr
   1988   *The Nature of Expertise.* Hillsdale, NJ: Erlbaum.
Church, G.J., and J. Greenwald
   1993   Jobs in an age of insecurity. *Time*, November 22.
Cohen, S., and D. Bailey
   1997   What makes teams work: Group effectiveness research from the shop floor to the executive suite. *Journal of Management* 23:239-290.
Commission on the Future of Worker Management Relations
   1994   *Fact Finding Report.* Washington, DC: U.S. Departments of Labor and Commerce.
   1995   *Final Report and Recommendations.* Washington, DC: U.S. Departments of Labor and Commerce.
The Conference Board
   1997   HR Executive Review: Implementing the New Employment Contract. Volume 4, Number 4, page 12. New York: The Conference Board.
Conk, M.A.
   1978   *The United States Census and Labor Force Change: A History of Occupational Statistics, 1870-1940.* Ann Arbor: University of Michigan Research Press.
Coombs, R.W.
   1984   Long-term trends in automation. Pp. 147-162 in *New Technology and the Future of Work and Skills*, P. Marstrand, ed. London: Francis Pinter.
Cotton, J.L.
   1993   *Employee Involvement: Methods for Improving Performance and Work Attitudes.* Newbury Park, CA: Sage.
Cowen, R.S.
   1983   *More Work for Mother: The Ironies of Household Technologies from the Open Hearth to the Microwave.* New York: Basic Books.

Cunningham, J.W.
   1988   Occupation analysis inventory. Pp. 975-90 in *The Job Analysis Handbook For Business, Industry, and Government*, Volume 2, S. Gael, ed. New York: Wiley.
Cunningham, J.W., R.R. Boese, R. Neeb, and J.J. Pass
   1983   Systematically derived work dimensions: Factor analysis of the Occupational Analysis Inventory. *Journal of Applied Psychology* 68:232-252.
Cunningham, J.W., T.C. Tuttle, J.R. Floyd, and J.A. Bates
   1971   *Occupation Analyis Inventory*. Center for Occupational Health. Raleigh: North Carolina State University.
Cunningham, J.W., W.E. Wimpee, and R.D. Ballentine
   1990   Some general dimensions of work among U.S. Air Force enlisted occupations. *Military Psychology* 2:33-45.
Cunningham, J.W., T.E. Powell, W.E. Wimpee, M.A. Wilson, and R.D. Ballentine
   1996   Ability-requirement factors for general job elements. *Military Psychology* 8:219-234.
Davis, J.A., and T.W. Smith
   1992   *The General Social Survey: A User's Guide*. Newbury Park, CA: Sage.
Dawis, R.V., and L.H. Lofquist
   1984   *A Psychological Theory of Work Adjustment: An Individual-Differences Model and Its Applications*. Minneapolis: University of Minnesota Press.
Day, G.
   1990   *Market Driven Strategy: Processes for Creating Value*. New York: Free Press.
Defense Manpower Data Center
   1998   Special Tabulations. Monterey, California.
Dempsey, R.E.
   1993   An occupational classification system for collecting employment data from both households and employers. Pp. 235-248 in *Proceedings of the International Occupational Classification Conference, Report 833*. Bureau of Labor Statistics. Washington, DC: U.S. Department of Labor.
DeNisi, A.S., E.T. Cornelius, III, and A.G. Blencoe
   1987   Further investigation of common knowledge effects on job analysis ratings. *Journal of Applied Psychology* 72:262-268.
Depuy, W.E.
   1995   For the joint specialist: Five steep hills to climb. From *Parameters* Summer:141-150. Internet addresss: http://carlisle-www.army.mil/usawc/Parameters/1995/depuy.htm.
Derber, C.
   1982   Toward a new theory of professionals as workers: Advanced capitalism and postindustrial labor. Pp.193-208 in *Professionals as Workers: Mental Labor in Advanced Capitalism*, C. Derber, ed. Boston: G. K. Hall and Co.
Derber, C., and W.A. Schwartz
   1991   New mandarins or new proletariat? Professional power at work. Pp. 71-96 in *Research in the Sociology of Organizations*, Volume 8, P.S. Tolbert and S.R. Barley, eds. Greenwich, CT: JAI Press.

Dertouzos, M.L., and J. Moses
  1979   *The Computer Age: A Twenty Year View.* Cambridge, MA: MIT Press.
Diebold, F.X., D. Neumark, and D. Polsky
  1996   Comment on Kenneth A. Swinnerton and Howard Wial, "Is job stability declining in the U.S. economy? *Industrial and Labor Relations Review* 49(2):348-352.
  1997   Job stability in the United States. *Journal of Labor Economics* 15(2):206-233.
DiNardo, J.E., and J.-S. Pischke
  1997   The returns to computer use revisited: Have pencils changed the wage structure too? *The Quarterly Journal of Economics* 112:291-303.
DiNardo, J., B. Fortin, and T. Lemieux
  1996   Labor market institutions and the distribution of wages, 1973-1992: A semi-parametric analysis. *Econometrica* 64:1001-1044.
Donnelly, R.G., and D.S.Kezsbom
  1994   Overcoming the responsibility-authority gap: An investigation of effective project team leadership for a new decade. *Cost Engineering* 36(5):33-41.
Dunlop, J.T.
  1976   The limits of legal compulsion. *Labor Law Journal* 27:67.
Dunlop, J.T., and D. Weil
  1996   Diffusion and performance of modular production in the U.S. apparel industry. *Industrial Relations* 35:334-354.
Eaton, S.
  1997   *Pennsylvania's Nursing Homes: Promoting Quality Care and Quality Jobs.* Harrisburg, PA: Keystone Research Center.
*Economic Report of the President*
  1995   *The Economic Report of the President.* Washington, DC: U.S. Government Printing Office.
Elias, P.
  1993   Developing an occupational classification system for the European community: Efforts to harmonize national systems. Pp. 295-298 in *Proceedings of the International Occupational Classification Conference, Report 833.* Washington, DC: U.S. Department of Labor.
England, P.
  1981   Assessing trends in occupational sex segregation, 1900-1976. In *Sociological Perspectives on Labor Markets*, I. Berg, ed. New York: Academic Press.
Farber, H.S.
  1995   *Are Lifetime Jobs Disappearing?* Working Paper. Cambridge, MA: National Bureau of Economic Research.
  1996   The Changing Face of Job Loss in the United States, 1981-1995. Working Paper 360, Industrial Relations Section, Princeton University, Princeton, NJ.
  1997   The changing face of job loss. In *Beyond Downsizing: Staffing and Workforce Management for the Milennium.* Woodcliff Lake, NJ: Lee Hecht Harrison.

Faulkner, R.S.
  1983  *Music on Demand: Composers and Careers in the Hollywood Film Industry.* New Brunswick, NJ: Transaction Books.
Ferman, L., M. Hoyman, J. Cutcher-Gershenfeld, and E.J. Savoie
  1990  *New Developments in Worker Training: A Legacy for the 1990s.* Madison, WI: Industrial Relations Research Association.
Fine, C.H.
  1998  *Clockspeed: Winning Industry Control in the Age of Temporary Advantage.* Reading, MA: Perseus Books, 1998.
Fine, L.
  1990  *The Souls of the Skyscraper: Female Clerical Workers in Chicago, 1870-1930.* Philadelphia: Temple University Press.
Fine, S.A., and W.W. Wiley
  1971  *An Introduction to Functional Job Analysis.* Kalamazoo, MI: The W.E. Upjohn Institute for Employment Research.
Fleishman, E.A.
  1967  Development of a behavior taxonomy for describing human tasks: A correlational-experimental approach. *Journal of Applied Psychology* 51:1-10.
  1992  *Fleishman Job Analysis Survey (F-JAS).* Bethesda, MD: Management Research Institute.
Fleishman, E.A., and M.D. Mumford
  1988  The ability rating scales. Pp. 917-935 in *Handbook of Job Analysis for Business, Industry, and Government,* S. Gael, ed. New York: Wiley.
  1989  Individual attributes and training performance: Applications of abilities taxonomies in instructional systems design. Pp. 183-255 in *Frontiers of Industrial and Organizational Psychology,* Volume 3, L. Goldstein, ed. San Francisco: Jossey-Bass.
Fleishman, E.A., and M.K. Quaintance
  1984  *Taxonomies of Human Performance: The Description of Human Tasks.* New York: Academic Press.
Fleishman, E.A., and M.E. Reilly
  1992  *Handbook of Human Abilities: Definitions, Measurement, and Job Task Requirements.* Palo Alto, CA: Consulting Psychologists Press.
Fligstein, N.
  1990  *The Transformation of Corporate Control.* Cambridge, MA: Harvard University Press.
Floyd, S., and W. Wooldridge
  1996  *The Strategic Middle Manager.* San Francisco: Jossey-Bass.
Foner, A., K. Schwab
  1983  Work and retirement in a changing society. In *Aging in Society: Selected Reviews of Recent Research,* M.W. Riley, B.B. Hess, and K. Bond, eds.. Hillsdale, NJ: Lawrence Erlbaum Associates.
Force XXI Operations
  1994  Force XXI Operations: A Concept for the Evolution of Full-Dimensional Operations for the Strategic Army of the Early Twenty-First Century. TRADOC Pamphlet 525-5.

*Fortune*
1994    The new deal: What companies and employees owe one another. *Fortune.* June 13, p. 44.

Fortin, N., and T. Lemieux
1997    Institutional changes and rising wage inequality: Is there a linkage? *Journal of Economic Perspectives* 11:75-96.

Freidson, E.
1973    Professions and the occupational principle. Pp. 19-37 in *Professions and Their Prospects*, E. Friedson, ed. Beverly Hills: Sage.

Gallie, D.
1978    *In Search of the New Working Class: Automation and Social Integration Within the Capitalist Enterprise.* Cambridge, MA: Cambridge University Press.
1994    Patterns of skill change: Upskilling, deskilling or polarization? In *Skill and Occupational Change*, R. Penn, M. Rose, and J. Rubery, eds. London: Oxford University Press.

Garson, B.
1975    *All the Livelong Day: The Meaning and Demeaning of Routine Work.* New York: Doubleday.
1988    *The Electronic Sweatshop: How Computers Are Transforming the Office of the Future in the Factory of the Past.* New York: Simon and Shuster.

George, W.R., and C. Marshall, eds.
1985    *New Services.* Chicago: American Marketing Association.

Gereffi, G.
1994    The international economy and economic development. Pp. 206-233 in *The Handbook of Economic Sociology*, N.J. Smelser and R. Swedberg, eds. New York: Russell Sage.

Gill, R.T., N. Glazer, and S. Thernstrom
1992    *Our Changing Population.* Englewood Cliffs, NJ: Prentice-Hall.

Gillen, M.
1994    *Models of Management.* Chicago: University of Chicago Press.

Gittleman, M., M. Horrigan, and M. Joyce
1998    "Flexible" workplace practices: Evidence from a nationally representative sample. *Industrial and Labor Relations Review* 52(1) October:99-115.

Gorden, D.
1994    Bosses of a Different Stripe: Monitoring and Supervision Across Advanced Economies. Working paper #49, January. Faculty of Economics, New School for Social Research.

Gouldner, A.W.
1960    The norm of reciprocity: A preliminary statement. *American Sociological Review* 25:161-178.

Graham, L.
1995    *On the Line at Subaru-Isuzu.* Ithaca, NY: ILR Press.

Granovetter, M.
1995    *Getting a Job: A Study of Contacts and Careers, Second Edition* (original edition 1974). Chicago: University of Chicago Press.

Gregory, D.J., and R.K. Park
1992    *Occupational Study of Federal Executives, Managers, and Supervisors: An Application of the Multipurpose Occupational Systems Analysis Inventory—Closed Ended (MOSAIC).* Personnel Resources and Development Center. Washington, DC: U.S. Office of Personnel Management.

Grissmer, D.W., S.N. Kirby, R.J. Buddin, J.H. Kawata, J.M. Sollinger, and S. Williamson
1994    *Prior Service Personnel: A Potential Constraint on Increasing Reliance on Reserve Forces.* RAND/RB-7501. National Defense Research Institute. Santa Monica, CA: The RAND Corporation.

Guest, D.
1997    Human resource management and performance: A review and research agenda. *International Journal of Human Resource Management* 8:263-276.

Gutek, B.
1995    *The Dynamics of Service: Reflections on the Changing Nature of Customer/Provider Interactions.* San Francisco: Jossey-Bass.

Hackman, R.J.
1987    The design of work teams. Pp. 315-342 in *Handbook of Organizational Behavior*, J.W. Lorsch, ed. Englewood Cliffs, NJ: Prentice-Hall.

Hackman, R.J., and G. Oldham
1975    Development of the job diagnostic survey. *Journal of Applied Psychology* 60:159-170.

Hamermesh, D.S.
1996    The Timing of Work Time Over Time. NBER Working Paper No. 5855.

Handy, C.
1989    *The Age of Unreason.* Boston: Harvard University Press.

Harrison, B.
1994    *Lean and Mean: The Changing Landscape of Corporate Power in the Age of Flexibility.* New York: Basic Books.

Harvey, R.J.
1991    Research Monograph: The Development of the Common-Metric Questionnaire (CMQ). Personnel Systems and Technologies Corporation and Virginia Polytechnic Institute and State University. Available from author.

Hayward, M.D., and W.R. Grady
1990    Work and retirement among a cohort of older men in the United States, 1966-1983. *Demography* 27:337-356.

Herzenberg, S., et al.
1998    *New Rules for a New Economy: Employment and Opportunity in Post-industrial America.* Ithaca, NY: Cornell University ILR Press.

Heskett, J., W.E. Sasser, Jr., and C. Hart
1990    *Service Breakthroughs: Changing the Rules of the Game.* New York: Free Press.

Hill, M.S.
1992    *The Panel Study of Income Dynamics: A User's Guide.* Newbury Park, CA: Sage.

Hirschhorn, L.
    1984  *Beyond Mechanization*. Cambridge, MA: MIT Press.
Hochschild, A.R.
    1983  *The Managed Heart: Commercialization of Human Feeling*. Berkeley: University of California Press.
    1997  *The Time Bind: When Work Becomes Home and Home Becomes Work*. New York: Metropolitan Books.
Hodes, B.
    1992  A new foundation in business culture: Managerial coaching. *Industrial Management* 34(5):27-28.
Holland, J.
    1985  *Making Vocational Choices: A Theory of Careers (Second Edition)*. Upper Saddle River, NJ: Prentice Hall.
Holmes, H.A.
    1996  Civil Affairs Soldiers are Crucial to Peace. Prepared remarks for the Worldwide Civil Affairs Conference, Washington, June 20. *Defense Issues* 11(60). Internet address: http://www.defenselink.mil/pubs/di96/di1160.html.
    1997  Military Operations in the Post-Cold War Era. Prepared remarks at the Intelligence in Partnership Conference, Joint Military Intelligence College, Andrews Air Force Base, MD, June 26. *Defense Issues* 12(34). Internet address: http://www.defenselink.mil/pubs/di97/di1234.html.
Hounshell, D.
    1984  *From the American System to Mass Production: 1800-1932*. Baltimore: Johns Hopkins Press.
Houseman, S.N.
    1997  Temporary, Part-Time, and Contract Employment in the United States: New Evidence from an Employer Survey. Unpublished manuscript, W.E. Upjohn Institute for Employment Research.
Houseman, S.N., and A.E. Polivka
    1998  The Implications of Flexible Staffing Arrangements for Job Security. Paper presented at conference on Changes in Job Stability and Job Security, sponsored by the Russell Sage Foundation, New York.
Hughes, E.C.
    1958  *Men and Their Work*. Glencoe, IL: Free Press.
Hughes, T.P.
    1983  *Networks of Power: Electrification in Western Society, 1880-1930*. Baltimore: Johns Hopkins Press.
Hunter, L.W.
    1998a  Transforming retail banking: Inclusion and segmentation in service work. In *Employment Strategies*, P. Cappelli, ed. New York: Oxford University Press. Forthcoming.
    1998b  Customer differentiation, institutional fields, and the quality of entry level jobs. Article under review by *Industrial and Labor Relations Review*.

Hunter, L., and J. Lafkas
  1998   Information technology, work practices, and wages. *In Proceedings of the 50th annual meeting of the Industrial Relations Research Association.* Madison, WI: Industrial Relations Research Association.
Huntington, S.P.
  1959   *The Soldier and the State.* Cambridge, MA: The Belknap Press of Harvard University Press.
Hutzel, T., and G. Varney
  1992   The supervisor's role in self-directed workteams. *Journal for Quality and Participation* 15(7):36-41.
Ichniowski, C., K. Shaw, and G. Prennushi
  1997   The effects of human resource practices on productivity. *American Economic Review* 87(3):291-314.
International Labour Office
  1990   *International Standard Classification of Occupations: ISCO-88.* Geneva, Switzerland: International Labour Organization.
Jackson, D., and J. Humble
  1994   Middle managers: New purpose, new directions. *Journal of Management Development* 13(3):15-21.
Jacobs, J.A.
  1998   Measuring time at work: Are self-reports accurate? *Monthly Labor Review* December:42-53.
Jacobs, J.A., and K. Gerson
  1998   Who are the overworked Americans? *Review of Social Economy* 56(4)Winter:442-459.
Jacoby, S.M.
  1985   *Employing Bureaucracy: Managers, Unions, and the Transformation of Work in American Industry, 1900-1945.* New York: Columbia University Press.
  1991   *Masters to Managers: Historical and Comparative Perspectives on American Employers.* New York: Columbia University Press.
Jaikumar, R.
  1986   Postindustrial manufacturing. *Harvard Business Review* (November-December):69-76.
Joint Chiefs of Staff
  1997   *Concept for Future Joint Operations: Expanding Joint Vision 2010.* Office of Primary Responsibility: Commander, Joint Warfighting Center, Building 96 Fenwick Road, Fort Monroe, VA 22361-5000.
Jones, B.
  1982   Destruction or redistribution of engineering skills? The case of numerical control. Pp. 179-200 in *The Degradation of Work: Skill, Deskilling and the Labour Process,* S.J. Wood, ed. London: Hutchinson.
Kageff, L.L., and J.H. Laurence
  1994   Test score trends and the recruit quality queue. In *Marching Toward the 21st Century,* M.J. Eitelberg and S.L. Mehay, eds. Westport, CT: Greenwood Press.

Kalleberg, A.L., E. Rasell, N. Cassirer, B.F. Reskin, K. Hudson, D. Webster, E. Appelbaum, and R.M. Spalter-Roth
   1997   Nonstandard Work, Substandard Jobs: Flexible Work Arrangements in the U.S. Washington, DC: Economic Policy Institute.
Kaminski, P.G.
   1996   Sustaining Flight Through Knowledge. Remarks at the Ira C. Eaker Distinguished Lecture on National Defense Policy, U.S. Air Force Academy, Colorado Springs, May 2. Defense Issues 11(42). Internet address: http://www.defenselink.mil/pubs/di96/di1142.html.
Katz, H.C., T.A. Kochan, and J.H. Keefe
   1987   Industrial relations and productivity in the U.S. automobile industry. Brookings Papers on Economic Activity 3:685-715.
Katz, L.F., and K.M. Murphy
   1992   Changes in relative wages, 1963-1987: Supply and demand factors. Quarterly Journal of Economics 107:35-78.
Keefe, J.H.
   1991   Numerically controlled machine tools and worker skills. Industrial and Labor Relations Review 44:503-519.
Keefe, J., and R. Batt
   1997   The United States. Chapter 1 in Telecommunications Restructuring: Work and Employment Relations Worldwide. Ithaca, NY: Cornell University Press.
Kelley, M.R.
   1986   Programmable automation and the skill question: A reinterpretation of the cross-national evidence. Human Systems Management 6:223-241.
Keltner, B., and D. Finegold
   1996   Adding value in banking: Human resource innovations for service firms. Sloan Management Review 38(1) Fall.
Keltner, B., and B. Jenson
   1998   Strategic Segmentation. Paper prepared for the The Changing Service Workplace conference. Wharton, University of Pennsylvania, October 16-17.
Kern, H., and M. Schumman
   1992   New Concepts of Production and the Emergence of the Systems Controller. Pp. 111-148 in Technology and the Future of Work, P.S. Adler, ed. New York: Oxford University Press.
Klein, J.
   1989   A reexamination of autonomy in light of new manufacturing practices. Human Relations 44:21-38.
   1994   Maintaining expertise in multiskilled teams. Pp. 145-166 in Advances in Interdisciplinary Studies of Work Teams, M.M. Beyerlein and D.A. Johnson, eds. Greenwich, CT: JAI Press.
Kochan, T.A., H.C. Katz, and R.B. McKersie
   1986   The Transformation of American Industrial Relations. New York: Basic Books.

Kochan, T.A., and P. Osterman
1994    *The Mutual Gains Enterprise.* Boston: Harvard Business School Press.
Kohl, G.
1993    Information technology and labor: A case study of telephone operators. *Workplace Topics* 3(1):101-111.
Kohn, M.L., and K.M. Slomczynski
1990    *Social Structure and Self-Direction: A Comparative Study of the United States and Poland.* Cambridge, MA: Basil Blackwell.
Krueger, A.B.
1993    How computers have changed the wage structure. *Quarterly Journal of Economics* 108:33-60.
Kuhn, S.
1989    The limits to industrialization: Computer software development in a large commercial bank. Pp. 266-78 in *The Transformation of Work?: Skill, Flexibility and the Labour Process,* S.J. Wood, ed. London: Unwin Hyman.
Kurtz, R., and G.K. Walker
1975    Test and pay technicians—Upward mobility in personnel. *Public Personnel Management* (July-August):259-62.
Lakewood Research
1988    Training Magazine's industry report 1988. *Training* (October):31-60.
1991    Industry report 1991. *Training* (October):31-59.
1997    Industry report 1997. *Training* (October):33-75.
Landes, David S.
1969    *The Unbound Prometheus: Technological Change and Industrial Development in Western Europe from 1750 to the Present.* Cambridge, England: Cambridge University Press.
Latour, B., and S. Woolgar
1979    *Laboratory Life: The Construction of Scientific Facts.* Princeton, NJ: Princeton University Press.
Lazonick, W.
1992    Controlling the market for corporate control: The historical significance of managerial capitalism. *Industrial and Corporate Change* 1:445-488.
Ledford, G.E., E.E. Lawler III, and S.A. Mohrman
1992    *Creating High Performance Organizations: Practices and Results of Employee Involvement and Total Quality Management in Fortune 1000 Companies.* San Francisco: Jossey Bass.
Lee Hecht Harrison
1997    *Beyond Downsizing: Staffing and Workforce Management for the Millennium.* Woodcliff Lake, NJ: Lee Hecht Harrison.
Leidner, R.
1993    *Fast Food, Fast Talk: Service Work and the Routinization of Everyday Life.* Berkeley: University of California Press.
Leontif, W.W.
1982    The distribution of work and income. *Scientific American* 247:188-204.
Levin, H.M., R.W. Rumberger, and C. Finnan
1990    Escalating Skill Requirements or Different Skill Requirements? *Confer-*

*ence on Changing Occuaptional Skill Requirements: Gathering and Assessing the Evidence.* Brown University, Providence, RI.

Levitt, T.
  1972   Production line approach to services. *Harvard Business Review* 50(5): 41-50.

Levy, F., and R.J. Murnane
  1996   With What Skills are Computers a Complement? Industrial Performance Center Working Paper, 1-29, Massachusetts Institute of Technology, Cambridge, MA.

Lewis, J.D.
  1995   *Connected Corporation: How Leading Companies Win Through Customer-Supplier Alliances.* New York: Free Press.

Lincoln, J.R., and A.L. Kalleberg
  1990   *Culture, Control, and Commitment: A Study of Work Organization and Work Attitudes in the United States and Japan.* Cambridge, England: Cambridge University Press.

Lipsky, D.B., and C.B. Donn, eds.
  1987   Airline industrial relations after deregulation. In *Collective Bargaining in American History.* Boston, MA: D.C. Health and Co.

Lorange, P., and J. Roos
  1992   *Strategic Alliances: Formation, Implementation, and Evolution.* New York: Oxford University Press.

Lovelock, C.
  1990   Managing interactions between operations and marketing and their impact on customers. Pp. 343-368 in *Service Management Effectiveness: Balancing Strategy, Organization and Human Resources, Operations, and Marketing*, D. Bowen, R.B. Chase, and T. Cummings, eds. Oxford, England: Jossey-Bass.

Lowe, G.
  1987   *Women in the Administrative Revolution: The Feminization of Clerical Work.* Cambridge, MA: Polity Press.

MacDonald, C.L., and C. Sirianni, eds.
  1996   *Working in the Service Sector.* Philadelphia: Temple University Press.

MacDuffie, J.P.
  1996   Automotive white collar: The changing status and roles of salaried employees in the North American auto industry. Pp. 81-126 in *Broken Ladders: Managerial Careers in the New Economy*, P. Osterman, ed. New York: Oxford University Press.

MacDuffie, J.P., and F. Pil
  1997   Changes in auto industry employment practices: An international overview. Pp. 9-44 in *After Lean Production*, T.A. Kochan, R.D. Lansbury, and J.P. MacDuffie, eds. Ithaca, NY: Cornell University Press.

Madden, R.C., and S. Tam
  1993   Australian Standard Classification of Occupations conceptual framework and directions for the second edition. Pp. 254-268 in *Proceedings*

*of the International Occupational Classification Conference.* Bureau of Labor Statistics Report 833. Washington, DC: U.S. Department of Labor.

Mankin, D.A., S. Cohen, and T. Bikson
1996 *Teams and Technology: Fulfilling the Promise of New Organization.* Boston: Harvard Business School Press.

Manz, C.C., D.E. Keating, and A. Donnellon
1990 Preparing for an organizational change to employee self-management: The management transition. *Organizational Dynamics* 19(2):15-26.

Marcotte, D.E.
1996 Has Job Stability Declined? Evidence from the Panel Study of Income Dynamics. Unpublished paper, Northern Illinois University.

McCormick, D.
1998 *The Downsized Warrior: America's Army in Transition.* New York: New York University Press.

McCormick, E.J.
1979 *Job Analysis: Methods and Applications.* New York: Amacom.

McCormick, E.J., P.R. Jeanneret, and R.C. Mecham
1969 *The Development and Background of the Position Analysis Questionnaire (PAQ).* Technical Report No. 5. Occupational Research Center. Lafayette, IN: Purdue University.
1972 A study of job characteristics and job dimensions as based on the Position Analysis Questionnaire (PAQ). *Journal of Applied Psychology (Monograph)* 56:347-368.

McKinsey Global Institute
1992 *Service Sector Productivity.* Washington, DC: McKinsey Global Institute.

Milgrom, P., and J. Roberts
1992 *Economics, Organizations, and Management.* Englewood Cliffs, NJ: Prentice-Hall.

Milligan, P.A.
1996 Regaining commitment. In *The New Deal in Employment Relationships: A Council Report.* Report No. 1162-96-CR. New York: The Conference Board.

Mills, C.W.
1951 *White Collar: The American Middle Class.* New York: Oxford University Press.

Mitchell, J.L., and W.E. Driskill
1996 Military job analysis: A historical perspective. *Military Psychology* 8(3):119-142.

*Monthly Labor Review*
1997 Table 4: Employment status of the population by sex, age, race, and Hispanic origin, monthly data seasonally adjusted. *Monthly Labor Review* July:61.

Morgan, B.B., A.S. Glickman, E.A. Woodard, A.S. Blaiwes, and E. Salas
1986 *Measurement of Team Behaviors in a Navy Environment.* Technical Report Number TR-86-014. Orlando, FL: Naval Training Systems Center.

Morse, D.
1969    *The Peripheral Worker*. New York: Columbia University Press.
Morse, N.C., and R. Weiss
1955    The function and meaning of work and the job. *American Sociological Review* 20:191-198.
Moskos, C.C.
1992    Armed Forces in the Post-Cold War Era, with Special Reference to the United States Army. Paper presented at the Workshop on Sociocultural Designs for the Future Army, University of Maryland, College Park.
Moskos, C.C., and F.R. Wood
1988    *The Military: More than Just a Job?* Washington, DC: Pergamon-Brassey's.
National Association of Staffing Services
1994    *Profile of the Temporary Workforce*. Alexandria, VA: National Association of Staffing Services.
National Center for Educational Statistics
1997    *The Condition of Education 1997*. Washington, DC: U.S. Department of Education.
National Occupational Information Coordinating Committee
1993    *National Occupational Information Coordinating Committee (NOICC) Master Crosswalk* (Version 4.0). [Electronic database]. Des Moines, IA: National Crosswalk Service Center.
National Research Council
1980    *Work, Jobs and Occupations: A Critical Review of the Dictionary of Occupational Titles*. A.R. Miller, D.J. Treiman, P.S. Cain, and P.A. Roos, eds. Committee on Occupational Classification and Analysis. Washington, DC: National Academy Press.
1986a   *Biotechnology: An Industry Comes of Age*. S. Olsen. Academy Industry Program. Washington, DC: National Academy Press.
1986b   *Computer Chips and Paper Clips: Technology and Women's Employment*. Vols. I and II. H. Hartmann, R. Kraut, and L. Tilly, eds. Committee on Women's Employment. Washington, DC: National Academy Press.
1994    *Information Technology in the Service Society*. Committee to Study the Impact of Information Technology on the Performance of Service. Washington, DC: National Academy Press.
1997a   *Enhancing Organizational Performance*. D. Druckman, J.E. Singer, and H. Van Cott, eds. Committee on Techniques for the Enhancement of Human Performance. Washington, DC: National Academy Press.
1997b   *Flight to the Future: Human Factors in Air Traffic Control*. C.D. Wickens, A.S. Mavor, and J.P. McGee, eds. Panel on Human Factors in Air Traffic Control Automation. Washington, DC: National Academy Press.
1998    *The Future of Air Traffic Control: Human Operators and Automation*. C.D. Wickens, A.S. Mavor, R. Parasuraman, and J.P. McGee, eds. Panel on Human Factors in Air Traffic Control Automation. Washington, DC: National Academy Press.

Negroponte, N.
  1995    *Being Digital.* New York: Knopf.
Nelsen, B., and S. Barley
  1993    *The Social Negotiation of a Recognized Occupational Identity.* Ithaca, NY: Cornell University.
Neumark, D., D. Polsky, and D. Hansen
  1997    Has Job Stability Declined Yet? New Evidence for the 1990s. NBER Working Paper No. 6330.
  In      Are Occupations Becoming Less Permanent? In *Job Stability and Job*
  Press   *Security.* New York: Russell Sage.
Nijhowne, S., and W. Silver
  1993    The standard occupational classification 1991: Statistical considerations in developing a classification of occupations. Pp. 299-319 in *Proceedings of the International Occupational Classification Conference.* Bureau of Labor Statistics Report 833. Washington, DC: U.S. Department of Labor.
9to5, Working Women Education Fund
  1992    *High Performance Office Work: Improving Jobs and Productivity: An Overview with Case Studies of Clerical Job Redesign.* Cleveland, OH: Working Women Education Fund.
Nocerra, J.
  1996    Living with layoffs. *Fortune* (April 1):69.
Nora, S., and A. Minc
  1981    *The Computerization of Society.* Cambridge, MA: MIT Press.
Norwood, S.
  1990    *Labor's Flaming Youth: Telephone Operators and Worker Militancy, 1878-1923.* Urbana and Chicago: University of Illinois Press.
Nye, D.E.
  1990    *Electrifying American: Social Meanings of a New Technology.* Cambridge, MA: MIT Press.
O'Brien, G.E.
  1992    Changing meanings of work. Pp. 44-66 in *Employment Relations: The Psychology of Influence and Control at Work,* J.F. Hartley and G.M. Stephenson, eds. Oxford, England: Basil Blackwell.
Office of Management and Budget
  1997    1997 Standard Occupational Classification revision. *Federal Register* 62:3638-36409.
Office of Technology Assessment
  1987    *International Competition in Services.* OTA-ITE-328. Washington, DC: U.S. Government Printing Office.
Olalla, J., and R. Echeverria
  1996    Management by coaching. *HR Focus* 73(1):16-18.
Olsten Corporation
  1994    *Skills for Success.* Melville, NY: Olsten Corporation.
Organization for Economic Cooperation and Development
  1997    OECD Working Papers. Occupational Classification (ISCO-88): Concepts, Methods, Reliability, Validity and Cross-National Comparability.

Volume V, Number 52. Paris: Organization for Economic Cooperation and Development.

Osborne, D.L.
    1997   Domestic trends to the year 2015: Forecasts from the Army 21 study. *Marching Toward the 21st Century: Military Manpower and Recruiting.* Westport, CT: Greenwood Press.

Osterman, P.
    1994   How common is workplace transformation and who adopts it? *Industrial and Labor Relations Review* 47:173-188.

    1996   *Broken Ladders: Managerial Careers in the New Economy.* New York: Oxford University Press.

Forth-   *Securing Prosperity: How the American Labor Market Has Changed and*
coming  *What to Do About It.* Princeton, NJ: Princeton University Press.

Paige, E., Jr.
    1996a  Striving for information superiority. Prepared remarks to the 311th Theater Signal Command Activation Dinner, Fort Meade, MD, June 22. *Defense Issues* 11(72). Internet address: http://www.defenselink.mil/pubs/di96/di1172.html.

    1996b  Ensuring joint force superiority in the information age. Prepared remarks at the Armed Forces Staff College, Norfolk, VA, July 30. *Defense Links* 11(82). Internet address: http://www.defenselink.mil/pubs/di96/di1182.html.

Passel, J.S., and B. Edmonston
    1992   Table 3. In *Immigration and Race: Recent Trends in Immigration to the United States.* Washington, DC: The Urban Institute.

Penn, R., M. Rose, and J. Rubery
    1994   *Skill and Occupational Change.* London: Oxford University Press.

Perrole, J.A.
    1986   Intellectual assembly lines: The rationalization of managerial, professional, and technical work. *Computers and the Social Sciences* 2:111-121.

Perrow, C.
    1967   A framework for the comparative analysis of organizations. *American Sociological Review* 32:194-208.

Perrucci, R.
    1971   Engineering: Professional servant of power. *American Behavioral Scientist* 14:492-505.

Peters, R.
    1994   The new warrior class. From *Parameters* Summer:16-26. Internet address: http://carlisle-www.army.mil/usawc/Parameters/1994/peters.htm.

Peterson, N.G., ed.
    1997   *Occupational Information Network (O\*NET) Research and Development.* Salt Lake City, UT: Utah Department of Employment Security.

Peterson, N.G., and P.R. Jeanneret
    1997   Job analysis: Overview and description of deductive methods. Pp. 13-

50 in *Applied Measurement Methods in Industrial Psychology*, D.L. Whetzel and G.R. Wheaton, eds. Palo Alto, CA: Davies-Black Publishing.

Peterson, N.G., M.D. Mumford, W.C. Borman, P.R. Jeanneret, and E.A. Fleishman, eds.

1995 *Development of Prototype Occupational Information Network (O*NET) Content Model* (Volumes 1 and 2). Salt Lake City, UT: Utah Department of Employment Security.

1999 *An Occupational Information System for the 21st Century: The Development of O*NET.* Washington, DC: American Psychological Association.

Peterson, N.G., M.D. Mumford, W.C. Borman, P.R. Jeanneret, E.A. Fleishman, and K.Y. Levin, eds.

1996 *O*NET Final Technical Report.* Salt Lake City, UT: Utah Department of Employment Security.

Pettigrew, A.M.

1973 *The Politics of Organizational Decision Making.* London: Tavistock.

Pine, J.

1993 *Mass Customization.* Boston: Harvard Business School Press.

Pink, D.H.

1998 Free agent nation. *Fast Company* December/January(12):131-47.

Piore, M.J., and C. Sabel

1984 *The Second Industrial Divide.* New York: Basic Books.

Polsky, D.

1996 Changes in the Consequences of Job Separations in the U.S. Economy. Working Paper, University of Pennsylvania.

Porter, M.

1992 Capital Choices: Changing the Way America Invests in Industry. Research report presented to the Council on Competitiveness and cosponsored by the Harvard Business School.

Price, D.J.D.S.

1986 *Little Science, Big Science ... and Beyond.* New York: Columbia University Press.

Primoff, E.S., and S.A. Fine

1988 A history of job analysis. Pp. 14-29 in *The Job Analysis Handbook for Business, Industry, and Government, Volume 1*, S. Gail, ed. New York: Wiley.

The Psychological Corporation

1993 *The Common Metric Questionnaire: A Job Analysis System.* San Antonio, TX: The Psychological Corporation.

Reichheld, F.F.

1996 *The Loyalty Effect.* Cambridge, MA: Harvard Business School Press.

Reimer, D.J., Chief of Staff

1997a *Army Vision 2010.* Internet address: aanjun96.htm at www-tradoc. monroe.army.mil.

1997b The Army and the cyberspace crossroads. Prepared remarks at the Armed Forces Communications and Electronic Association TECHNET

'97, Washington, DC, June 17. *Defense Issues* 12(33). Internet address: http://www.defenselink.mil/pubs/di97/di1233.html.

1997c   *The Annual Report on The Army After Next Project to the Chief of Staff of the Army, July 1997.* Washington, DC: U.S. Army.

Reskin, B.
1993    Sex segregation in the workplace. *Annual Review of Sociology* 19(Annual Reviews):241-270.

Rifkin, J.
1995    *The End of Work.* New York: Putnam.

Ritti, R.R.
1971    *The Engineer in the Industrial Corporation.* New York: Columbia University Press.

Ritzer, G.
1998    *The McDonaldization of Society: An Investigation into the Changing Character of Contemporary Social Life.* Thousand Oaks, CA: Pine Forge Press.

Roberts, M.
1993    The national occupation classification of Canada. Pp. 320-321 in *Proceedings of the International Occupational Classification Conference.* Bureau of Labor Statistics Report 833. Washington, DC: U.S. Department of Labor.

Robinson, J.P., and A. Bostrom
1994    The overestimated workweek? What time diary measures suggest. *Monthly Labor Review* 117(8):11-23.

Robinson, S.L.
1996    Trust and the breach of the psychological contract. *Administrative Science Quarterly* 41:574-599.

Robinson, S.L, M.S. Kraatz, and D.M. Rousseau
1994    Changing obligations and the psychological contract: A longitudinal study. *Academy of Management Journal* 37:137-152.

Roe, M.J.
1994    *Strong Managers, Weak Owners.* Princeton, NJ: Princeton University Press.

Rollins, J.
1985    *Between Women: Domestics and Their Employers.* Philadelphia: Temple University Press.

Rones, P.L., R.E. Ilg, and J.M. Gardner
1997    Trends in hours of work since the mid-1970's. *Monthly Labor Review* April:3-14.

Rose, S.J.
1995    Declining Job Security and the Professionalization of Opportunity. Research Report 95-04. National Commission for Employment Policy, Washington, DC.

Rose, A.M., B. Hesse, P.A. Silver, and J.S. Dumas, eds.
1996    *O*NET: An Informational System for the Workplace. Designing an Electronic Infrastructure.* Salt Lake City, UT: Utah Department of Employment Security.

1999    Database design and development: Designing an electronic infrastruc-
        ture. Pp. 273-288 in *An Occupational Information System for the 21st Cen-
        tury: The Development of O\*NET*, N.G. Peterson, M.D. Mumford, W.C.
        Borman, P.R. Jeanneret, and E.A. Fleishman, eds. Washington, DC:
        American Psychological Association.
Rosenbaum, J.
1984    *Career Mobility in a Corporate Hierarchy*. Orlando, FL: Academic Press.
Rousseau, D.
1995    *Psychological Contracts in Organizations*. Thousand Oaks, CA: Sage Pub-
        lications.
Rousseau, D., and R.J. Anton
1991    Fairness and obligations in termination decisions: The role of contribu-
        tions, promises, and performance. *Journal of Organizational Behavior*
        12:287-299.
Rubinstein, S.
Forth-  The impact of co-management on quality performance: The case of the
coming  Saturn Corporation. Forthcoming in *Industrial and Labor Relations Re-
        view*.
Russell, T.L., J.L. Crafts, F.A. Tagliareni, R.A. McCloy, and P. Barkley
1994    *Job Analysis of Special Forces Jobs*. Alexandria, VA: U.S. Army Research
        Institute for the Behavioral and Social Sciences.
Russell, T.L., M.D. Mumford, and N.G. Peterson
1995    *Applicability of the Department of Labor's O\*NET for Army Occupational
        Analysis: Final Report, 30 September 1995*. Alexandria, VA: U.S. Army
        Research Institute for the Behavioral and Social Sciences.
Rynes, S.L., M.O. Orlitzky, and R.D. Bretz, Jr.
1997    Experienced hiring versus college recruiting: Practices and emerging
        trends. *Personnel Psychology* 50(2):309-339.
Sanders, P.
1997    Training and test ranges: A 21st century partnership. Remarks at the
        National Training Systems Association and International Test and
        Evaluation Association Workshop, Norfolk, VA, November 18. *Defense
        Issues* 12(57). Internet address: http://www.defenselink.mil/pubs/
        di97/di1257.html.
Saville and Holdsworth Ltd. USA, Inc.
1990    *Work Profiling System Manual*. Boston: Saville and Holdsworth Ltd.
        USA, Inc.
Saxenion, A.
1994    *Regional Advantage: Culture and Competition in Silicon Valley and Route
        128*. Cambridge, MA: Harvard University Press.
Schacht, J.
1985    *The Making of Telephone Unionism 1920-1947*. New Brunswick, NJ:
        Rutgers University Press.
Schlesinger, L., and J. Heskett
1991    Breaking the cycle of failure in services. *Sloan Management Review*
        32(Spring):17-28.

Schmidt, S.R., and S.V. Svorny
  Forth-  Recent trends in job security and stability. *Journal of Labor Research.*
  coming

Schneider, B., and D.E. Bowen
  1995   *Winning the Service Game.* Cambridge, MA: Harvard Business School
         Press.

Schoomaker, General P.J.
  1998   Special operations forces: The way ahead. Statement presented to
         members of the U.S. Special Operations Command. *Defense Issues*
         13(10). Internet address: http://www.defenselink.mil/pubs/di98/
         di1310.html.

Schor, J.B.
  1992   *The Overworked American: The Unexpected Decline of Leisure.* New York:
         Basic Books.

Scott, E., P. Cappelli, and K.C. O'Shaughnessy
  1996   Management jobs in the insurance industry: Organizational deskilling
         and rising pay inequality. Pp. 126-154 in *Broken Ladders: Managerial
         Careers in the New Economy*, P. Osterman, ed. New York: Oxford Uni-
         versity Press.

Scott, W.R.
  1981   *Organizations: Rational, Natural, and Open Systems.* Englewood Cliffs,
         NJ: Prentice Hall.

Segal, D.R.
  1993   *Organizational Designs for the Future Army, Special Report.* Alexandria,
         VA: U.S. Army Research Institute for the Behavioral and Social Sci-
         ences.

Segal, L.M., and D.G. Sullivan
  1996   The growth of temporary services work. *Journal of Economic Perspec-
         tives* 11(2):117-136.

Sellman, W.S.
  1995   Since we are reinventing everything else, why not occupational analy-
         sis. Pp. 8-18 in *Proceedings of the Ninth International Occupational Analy-
         sis Workshop.* San Antonio, TX: USAF Occupational Measurement
         Squadron.

Senate Armed Services Committee
  1997   Excerpts from the Departments of Army, Navy and Air Force posture
         statements as presented to the Senate Armed Services Committee. Feb-
         ruary 25. *Defense Issues* 12(26). Internet address: http://www.
         defenselink.mil/pubs/di97/di1226.html

Shalikashvili, J.M., Chairman of the Joint Chiefs of Staff
  1996a  The world in 2010 will be as challenging as today's. Prepared remarks
         at the Council on Foreign Relations, New York, November 7. *Defense
         Issues* 11(101). Internet address: http://www.defenselink.mil/pubs/
         di96/di1101.html.
  1996b  *Joint Vision 2010: America's Military, Preparing for Tomorrow.* Washing-
         ton, DC: Chairman of the Joint Chiefs of Staff.

1997 *National Military Strategy: Shape, Respond, Prepare Now—A Military Strategy for a New Era*. Washington, DC: Chairman of the Joint Chiefs of Staff.

Shimada, H., and J.P. MacDuffie
1987 Industrial Relations and Humanware. Working Paper. MIT Sloan School of Management.

Silvestri, G.T.
1995 Occupational employment to 2005. *Monthly Labor Review* 118:60-87.
1997 Occupational employment projections to 2006. *Monthly Labor Review* November:58-83.

Smith, C.
1987 *Technical Workers: Class, Labour and Trade Unionism*. London: MacMillan Education Ltd.

SOC Federal Register Notice
1998 August 5:3. Internet address: http://www.bls.gov/soc/soc_text.htm.

Spangler, E.
1986 *Lawyers for Hire*. New Haven, CT: Yale University Press.

Spenner, K.I.
1979 Temporal changes in work content. *American Sociological Review* 44:968-975.
1983 Temporal change in the skill level of work. *American Sociological Review* 48:824-837.
1988 Occupations, work settings and the course of adult development: Tracing the implications of select historical changes. Pp. 243-285 in *Life-Span Development and Behavior (9)*, P.B. Baltes, D.L. Featherman, and R.M. Lerner, eds. Hillsdale, NJ: Lawrence Erlbaum Associates.
1990 Skill: Meanings, methods, and measures. *Work and Occupations* 17:399-421.
1995 Technological change, skill requirements, and education: The case for uncertainty. In *The New Modern Times: Factors Reshaping the World of Work*, D.B. Bills, ed. Albany, NY: SUNY Press.

Spenner, K.I., L.B. Otto, and V.R.A. Call
1980 *Estimates of Third Edition DOT Job Characteristics for 1970 Census Occupation-Industry Categories*. Omaha, NE: Boys Town Center.

Steinberg, R.J.
1990 Social construction of skill: Gender, power, and comparable worth. *Work and Occupations* 17:449-482.
1992 Gendered instructions: Cultural lag and gender bias in the Hay System of Job Evaluation. *Work and Occupations* 19:387-423.

Stewart, T.A.
1997 *Intellectual Capital: The New Wealth of Nations*. New York: Doubleday.

Stinchcombe, A.L.
1959 Bureaucratic and craft administration of production: A comparative study. *Administrative Science Quarterly* 4:168-187.

Sweet, J.A.
1973 *Women in the Labor Force*. New York: Seminar Press.

Swinnerton, K., and H. Wial
   1995   Is job stability declining in the U.S. economy? *Industrial and Labor Relations Review* 48(2)(January):293-304.

Tannenbaum, S.I., R.L. Beard, and E. Salas
   1992   Team building and its influence on team effectiveness: An examination of conceptual and empirical developments. In *Issues, Theory, and Research in Industrial/Organizational Psychology*, K. Kelley, ed. Amsterdam, Netherlands: Elsevier Science Publishers B.V.

Tannenbaum, S.I., E. Salas, and J.A. Cannon-Bowers
   1996   Promoting team effectiveness. Pp. 503-529 (Chapter 20) in *Handbook of Work Group Psychology*, M.A. West, ed. New York: Wiley.

Teitelman, R.
   1989   *Gene Dreams: Wall Street, Academia, and the Rise of Biotechnology.* New York: Basic.

Thompson, J.
   1967   *Organizations in Action: The Social Basis of Administrative Theory.* New York: McGraw-Hill.

Tilly, C.
   1996   *Half a Job: Bad and Good Part-Time Jobs in a Changing Labor Market.* Philadelphia: Temple University Press.

Timenes, N., Jr.
   1996   Managing the force drawdown. In *Professionals on the Front Line: Two Decades of the All-Volunteer Force*, J.E. Fredland, C.L.Gilroy, R. Little, and W.S. Sellman, eds. Washington, DC: Brassey's.

Ulrich, D.
   1997   Judge me more by my future than by my past. *Human Resources Management* 26:5-8.

U.S. Army
   1997   *America's Army ... Into the 21st Century.* Program Analysis and Evaluation Directorate. Washington, DC: U.S. Army.

U.S. Army Research Institute for the Behavioral and Social Sciences
   1996   *The Army's Occupational Analysis Program: Integrating Manpower, Personnel and Training Requirements.* Progress Report, November. Internet address: http://www-ari.army.mil/pdf/oap.pdf.

U.S. Army Training and Doctrine Command
   1996   Report on the Army After Next Project, June 1996. Internet address: aanjun96.htm at www-tradoc.monroe.army.mil.

U.S. Army War College
   1997   *How the Army Runs: A Senior Leader Reference Handbook, 1997-1998.* Carlisle Barracks, PA: U.S. Army War College.

U.S. Bureau of the Census
   1992   Table 1. *Current Population Reports* 1092:25. Washington, DC: U.S. Department of Commerce.
   1996b  United States Population Estimates, by Age, Sex, Race, and Hispanic Origin, 1990 to 1996, with Associated Updated Tables for Recent

Months. Release PPL-57. Population Division. Washington, DC: U.S. Bureau of the Census.

U.S. Department of Commerce
  1980  *Standard Occupational Classification Manual.* Washington, DC: U.S. Department of Commerce.
  1990a  *Classified Index of Occupations and Industries.* Bureau of the Census. Washington, DC: U.S. Department of Commerce.
  1990b  *Alphabetical Index of Industries and Occupations, 1990 Census of Population and Housing.* Bureau of the Census. Washington, DC: U.S. Department of Commerce.
  1992  Current Population Survey, *Industry and Occupations Coding Procedures Manual.* Bureau of the Census. Washington, DC: U.S. Department of Commerce.

U.S. Department of Defense
  1997  *Population Representation in the Military Services, Fiscal Year 1996.* December. Washington, DC: Office of the Assistant Secretary of Defense for Force Management Policy.

U.S. Department of Labor
  1972  *Handbook for Analyzing Jobs.* Washington, DC: U.S. Department of Labor. (Revised in 1991.)
  1991  *Dictionary of Occupational Titles.* Washington, DC: U.S. Department of Labor.
  1992  *BLS Handbook of Methods, Bulletin 2414.* Bureau of Labor Statistics. Washington, DC: U.S. Department of Labor.
  1995  *Report on the American Workforce.* Washington, DC: U.S. Department of Labor.

U.S. General Accounting Office
  1998  *Military Recruiting: DOD Could Improve Its Recruiter Selection and Incentive Systems.* GAO/NSIAD-98-58. Washington, DC: U.S. General Accounting Office.

Useem, M.
  1996  Corporate restructuring and the restructured world of senior management. Pp. 23-54 in *Broken Ladders: Managerial Careers in the New Economy,* P. Osterman, ed. New York: Oxford University Press.

Valletta, R.G.
  1996  Has job security in the U.S. declined? *Federal Reserve Bank of San Francisco Weekly Letter* (96-07) (February 16).

Weber, M.
  1968  *Economy and Society.* Berkeley, CA: University of California Press.

Whalley, P.
  1986  *The Social Production of Technical Work.* Albany, NY: State University of New York Press.
  1991  Negotiating the boundaries of engineering: Professionals, managers, and manual work. Pp. 191-215 in *Research in the Sociology of Organizations,* Volume 8, P.S. Tolbert and S.R. Barley, eds. Greenwich, CT: JAI Press.

White, I.
    1993    The United Kingdom's standard occupational classification. Pp. 290-
            294 in *Proceedings of the International Occupational Classification Confer-
            ence*. Bureau of Labor Statistics Report 833. Washington, DC: U.S.
            Department of Labor.
White, J.P.
    1996    Adapting to new world realities. *Defense Issues* 11:45. Internet address:
            http://www.defenselink.mil/pubs/di96/di1145.html.
Whitely, R.
    1991    *The Customer Driven Company: Moving from Talk to Action*. Reading,
            MA: Addison-Wesley.
Whyte, W.H.
    1956    *The Organization Man*. New York, NY: Simon and Schuster.
Wilensky, H.L.
    1964    The professionalization of everyone. *American Journal of Sociology*
            70:137-58.
Williams, K., and C. O'Reilly
    1998    Forty years of diversity research: A review. Pp. 77-140 in *Research in
            Organizational Behavior, Volume 20*, B.M. Staw and L.L. Cummings, eds.
            Greenwich, CT: JAI Press.
Williamson, O.E., M.L. Wachter, and J.E. Harris
    1975    Understanding the employment relation: The analysis of idiosyncratic
            exchange. *Bell Journal of Economics* (6):250-278.
Woody, B.
    1989    Black women in the emerging services economy. *Sex Roles* 21:45-67.
Wootton, B.H.
    1993    Innovations in occupational classification: International lessons for re-
            vising the United States Standard Occupational Classification system.
            Pp. 326-33 in *Proceedings of the International Occupational Classification
            Conference, Report 833*. Bureau of Labor Statistics. Washington, DC:
            U.S. Department of Labor.
    1997    Gender differences in occupational employment. *Monthly Labor Review*
            April:15-24.
Worrell, D.L., W.N. Davidson III, and V.M. Sharma
    1991    Layoff announcements and stockholder wealth. *Academy of Manage-
            ment Journal* 34:662-678.
Wyatt Company
    1993    *Wyatt's 1993 Survey of Restructuring—Best Practices in Corporate Restruc-
            turing*. New York: Wyatt Company.
    1995    Measuring change in the attitudes of the American workforce. In *Wyatt
            Work USA*. New York: Wyatt Company.
Yamagami, T.
    1987    The Survival Strategy for the U.S. Steel Industry. Unpublished Master's
            Thesis. Sloan School of Management, Massachusetts Institute of Tech-
            nology.

Yates, J.
1993   *Control Through Communication: The Rise of System in American Manage-ment.* Baltimore: Johns Hopkins University Press.

Zabusky, S.E., and S.R. Barley
1996   Redefining success: Ethnographic observations on the careers of tech-nicians. In *Broken Ladders,* P. Osterman, ed. New York: Oxford University Press.

Zemke, R., and D. Schaff
1989   *The Service Edge: 101 Companies that Profit from Customer Care.* New York: NAL Books.

Zimmerman, C., and J. Enell
1988   Service industries. Pp. 33.1-72 in *Juran's Quality Control Handbook,* Fourth Edition, J.M. Duran and Frank Gryna, eds. New York: McGraw Hill.

Zuboff, S.
1989   *In the Age of the Smart Machine.* New York: Basic Books.

Zussman, R.
1985   *Mechanics of the Middle Class.* Berkeley: University of California Press.

# Appendixes

# APPENDIX
# A

# Prototype Evaluation

An initial set of studies was conducted with the O*NET™ instruments. Data were collected from incumbents in about 30 occupations out of an initial targeted sample of 70 occupations. In addition, occupational analysts rated 1,122 occupations on a subset of O*NET™ descriptors using information from the *Dictionary of Occupational Titles* to guide their ratings.

Return rates were disappointing for the questionnaires mailed to incumbents. Though 60 percent of all incumbents who were given questionnaires completed them, many establishments that initially agreed to participate failed to do so (only 27 percent actually participated). A number of factors may have contributed to the disappointing participation rates, suggesting that future data collections would benefit from shorter questionnaires, less burden on each establishment, monetary or other incentives for points of contact within each establishment and for incumbents, and greater use of organizations willing to volunteer their services or showing interest in applying O*NET™ results. However, results for the reliability of the questionnaires, as measured by interrater reliability coefficients, were acceptable. In addition, the correlations between the mean ratings of incumbents and the mean ratings of analysts were sufficiently high to warrant the use of the analysts' ratings as an interim O*NET™ database. Tables A.1 and A.2 summarize these results.

A large number of analyses was conducted on the data generated by these two studies of the prototype O*NET™. These are reported in Peterson et al. (1996, 1999) and Peterson (1997). Within each O*NET™ domain, these analyses were aimed at evaluating the adequacy of the structure of the variables and scales used. A number of broader analyses were also conducted, aimed at identifying relationships between variables within and across O*NET™ domains, identifying larger groups or clusters of O*NET™ occupations, and linking O*NET™ variables to variables outside the O*NET™ model, in particular to individual assessment variables. These kinds of analyses began to integrate the large amount of available O*NET™ information about both job descriptors and occupations and began to examine the relationships of O*NET™ variables and occupations to some of the available external information.

## CROSS-DOMAIN ANALYSES

Several different analytical approaches were used in cross-domain analyses, and each provided a somewhat different perspective on the relationships between descriptors from the various O*NET™ content domains. In general, these results strongly support the construct validity of the O*NET™ descriptors across all content domains, providing some interesting insights concerning cross-domain relationships. All of the tests of a priori cross-domain hypotheses showed that when strong correlations between O*NET™ descriptor scores were expected, strong correlations were in fact obtained.

In general, work activities involving information and people had strong correlations with many cognitive ability and skill requirements. The achievement and other more cognitively oriented work styles were also strongly related to activities involving information and people, as well as to cognitive ability and skill requirements. Work styles involving interpersonal interactions were positively correlated with activities and environments involving working with others. These relationships were summarized in the cross-domain factor analysis results, in which the first

TABLE A.1  Incumbent Interrater Agreement Coefficients for Each Scale Type

| Questionnaire | Scale | $r_k$ | $r_{30}$ |
|---|---|---|---|
| Skills | Level | .79 | .93 |
| | Importance | .79 | .93 |
| | Job entry requirement | .60 | .83 |
| Knowledge | Level | .86 | .95 |
| | Importance | .85 | .94 |
| Training, education, licensure, | Instructional program | .78 | .92 |
| and experience | Educational subject area | .74 | .90 |
| | Licensure | .85 | .95 |
| | Experience | .79 | .93 |
| Generalized work activities | Level | .80 | .92 |
| | Importance | .78 | .92 |
| | Frequency | .74 | .90 |
| Work context | | .87 | .95 |
| Organizational context | Across occupations | .64 | .84 |
| | Across organizations | .45 | .79 |
| Abilities | Level | .82 | .93 |
| | Importance | .82 | .93 |
| Occupational values | | .60 | .82 |
| Work styles | Level | .70 | .88 |
| | Importance | .67 | .86 |

Note: $r_k$ is the observed interrater agreement coefficient; $r_{30}$ is the estimated interrater agreement coefficient for 30 raters.

TABLE A.2  Mean Correlations Between Incumbents' and Analysts' Ratings

| Questionnaire | Scale | $r_{ia}$ |
|---|---|---|
| Skills | Level | .74 |
| | Importance | .67 |
| Knowledges | Level | .65 |
| | Importance | .65 |
| Generalized work activities | Level | .71 |
| | Importance | .61 |
| | Frequency | .53 |
| Work context | | .64 |
| Abilities | Level | .70 |
| | Importance | .65 |

Note: $r_{ia}$ is the mean correlation between incumbent and analyst mean occupation ratings.

factor was defined by descriptors related to interpersonal and managerial activities, cognitive skill requirements, and achievement-related worker characteristics. Although activities involving working with information and working with people had generally similar patterns of correlations with descriptors from other domains, the differences in these patterns of correlations support the construct validity of these composites. For example, a composite of several generalized work activities, labeled "working with information," was more strongly related to technical skills and math ability, whereas a composite labeled "working with people" was more strongly related to the people-oriented work styles. Manual and physical work activities were correlated with technical skills and with psychomotor and physical ability requirements. Environmental factors from the work context domain also tended to be positively correlated with manual and physical activities and related worker requirements. In fact, manual and physical activities, physical and psychomotor abilities, and environmental factors defined the second factor in the cross-domain factor analysis.

In addition to obtaining the expected relationships across domains, the analyses also generally showed that constructs conceptually unrelated do not correlate. For example, physical and psychomotor ability requirements were not significantly correlated with work activities involving information or people. These analyses also uncovered some conceptually interesting relationships. For example, office activities and related requirements tended to be negatively correlated with physical activities and related worker requirements. Technical skill requirements were negatively related with a work context that involves interacting with the public, and finally, law enforcement knowledge requirements correlated significantly with ability requirements such as psychomotor, vision and hearing, and spatial.

## GROUPING OCCUPATIONS

One advantage of the O*NET™ system is that occupations can be grouped on the basis of a variety of job descriptors, depending on the needs of the particular project. For example, occupations could be grouped on the basis of their scores on the ability requirements if the desire is to find groups of occupations that have similar ability profiles, say, for an entry-level employee selection program. Or occupations could be grouped on the basis of their generalized work activity scores if the goal is to identify occupations that tend to require that similar kinds of tasks are completed.

Given that the preferred method of clustering occupations is to examine the O*NET™ to form clusters that best meet the applied purpose, there remains some interest in identifying a set of occupational groups that can be used in a more general sense, i.e., the identification of a relatively small set of occupational groups that have been formed on the basis of a diverse set of the O*NET™ variables. These groups could be used in a general descriptive way, or they could be used as entry points for persons interested in exploring the world of work.

Toward that goal, several cluster analyses of O*NET™ occupations were completed. The first set of investigations focused on methodological variations and the statistical properties of different solutions, serving to guide a second set of more substantively oriented investigations. The central objective of this work was to evaluate the interpretability of occupational clusters generated using the occupational analyst ratings of knowledge, skill, ability, generalized work activities, and selected work context requirements for the 1,122 occupational units. The Ward-Hook and Q-factor analysis hybrid methods were initially used to generate solutions that had 50 occupational clusters. The Q-factor method provided the more interpretable solution, and accordingly, this method was employed to generate a more differentiated, 60-cluster solution, and a more parsimonious, 40-cluster solution. Although more research is obviously needed to evaluate the interpretability of a wider range of cluster solutions beyond those identified in these studies, these three solutions provide an initial set of occupational groups for descriptive and exploratory purposes.

## AGGREGATION OF DESCRIPTOR VARIABLES

The O*NET™ descriptor variables are hierarchically organized within each of the major variable types; for example, the 46 skill variables were originally organized into a smaller number of higher-order aggregates of those skills, based on the available theoretical and empirical evidence. These higher-level organizations were evaluated using the incumbent and analyst data. This was accomplished by computing correlations among the descriptors within each type (e.g., skills), factor analyzing the correlation matrices, and comparing the results to the original organizations of descriptors. The resulting reorganizations of descriptors were called "rational/empirical models" since they combined rational and empirical analyses to arrive at the structures.

The alternative rational/empirical models based on analyst data, not too surprisingly, appeared to describe the underlying structure of the analyst or transitional database better than the original content models for the ability, skill, and generalized work activity descriptors, respectively. Thus, these hierarchical structures are probably most appropriate for many uses of the transitional analyst dataset, especially for purposes of developing occupational scores for higher-level aggregates of descriptors. Regarding these aggregates, the second-order categories (for example, there are 16 second-order skill categories) were thought to be more appropriate for the development of aggregates than are the highest-order categories (for example, there are 6 highest-order skill categories). These latter categories are extremely broad, and aggregates formed at this level would lose too much information for most purposes. Several methods of combining descriptor scores into the higher level scores were investigated.

## LINKING O*NET™ JOB ANALYSIS INFORMATION TO THE ASSESSMENT OF JOB REQUIREMENTS

It is well known that professional and legal guidelines stipulate that the use of selection tests should be based on job analysis information. Consequently, the use of information obtained from job analyses is a critical element for identifying and establishing requirements for jobs, and the identification of important work behaviors and the employee characteristics that underlie those behaviors leads to the choice of appropriate selection tests. Research was conducted to demonstrate the applicability of the

O*NET™ database for the identification of assessment instruments or selection techniques to use when measuring aptitude requirements associated with selecting and placing employees. The design used to investigate the relationship between the O*NET™ job analysis results and potential assessment variables was predicated on the job component validation (JCV) model (McCormick, 1979).

Job component validation is one method to identify potential selection tests in situations in which it is not feasible to conduct other types of validation studies, primarily because of a lack of sufficient numbers of employees in the occupations for which selection procedures are to be developed. JCV involves two main hypotheses: (1) if various jobs have a given component in common, the attributes needed to fulfill the requirements of that component would be the same across the various jobs; and (2) the validity of a predictor of a job requirement defined by a job component is consistent across jobs.

The first step in the JCV process is the development and use of an objective job analysis procedure to document critical information about work behaviors and required worker characteristics for the job or occupation in question. Next, the JCV process examines the relationships between these specific job and worker characteristics and well-defined aptitude and ability characteristics. It was hypothesized that the O*NET™ data—in particular, the occupational analyst data—could be used as a source of job analysis information in the JCV process. The Position Analysis Questionnaire (PAQ) database, together with the O*NET™ analyst database, was used to see whether O*NET™ information could be used to accurately predict the General Aptitude Test Battery (GATB) and estimates of the Wonderlic test scores contained in the PAQ database. Using 249 occupations, a generally high level of accuracy was obtained in predicting these scores (e.g., cross-validated multiple correlation coefficients of .88 for predicting verbal aptitude, .82 for clerical perception, .64 for manual dexterity, and .81 for the Wonderlic). In addition, generalized work activities rationally linked to PAQ dimensions produced multiple correlations of similar magnitude as those generalized work activities empirically selected through a cross-validated regression analysis. Building on this research, it would seem plausible that the systematic and standardized job analysis information from the O*NET™ could be incorporated into a job component validation process that could assist organizations in identifying and selecting assessment systems for hiring and placement.

# APPENDIX
# B
# Current Occupational
# Analysis Systems

T his appendix describes current occupational analysis systems in more detail than is provided in the main text. Category/enumerative systems are presented first followed by descriptions of six illustrative descriptive analytic systems.

## CATEGORY/ENUMERATIVE SYSTEMS

### ISCO-88

The ISCO system uses two key concepts: job and skill. Job is defined as "a set of tasks and duties executed, or meant to be executed, by one person." Skill is defined as "the ability to carry out the tasks and duties of a given job." Operationally, four levels of skill are defined, entirely in terms of achieved education. The lowest level approximates primary school (about sixth grade in United States), the second approximates secondary school (about the twelfth grade in United States) but includes apprenticeships, the third approximates college education but not obtaining a degree, and the fourth includes undergraduate and graduate college education (International Labour Office, 1990:2-3).

The ISCO-88 structure is hierarchical, with 10 major groups at the top, 28 submajor groups, 116 minor groups, and 390 unit groups. Eight of the 10 major groups are categorized at one of the four skill levels ("armed forces" and "legislators, senior officials and managers," were not so categorized). For example, all occupations in the major group "clerks" are categorized at the second skill level, and all those in "elementary occupations" are at the first level. All descriptions are verbal, and no quantitative data are provided. Each major group, submajor group, and minor group is described by a general duty description and a list of tasks (usually no more than a brief paragraph in length). The lowest level "unit groups" also include names of "example occupations" and related occupations, in addition to the general duty and task list descriptions.

This development of this structure was "carried out in line with the recommendations and decisions of the Thirteenth and Fourteenth International Conferences of Labour Statisticians, held at the International Labour Office, Geneva, in 1982 and 1987" (International Labour Office, 1990:1). The underlying source data consist of population censuses, statistical surveys, and administrative records maintained at the national level.

The Organization for Economic Cooperation and Development (1997) recently reviewed the use of ISCO-88 in Europe and elsewhere around the world. It reached a number of conclusions:

- ISCO-88 has superseded ISCO-68 and has become the model for new national classifications in many countries, even those with previously existing systems.
- Levels of reliability of classifying occupations into ISCO categories remain fairly low, at 75 percent for the most detailed levels of categorization (about 350 categories).
- Aggregating to higher levels of categorization improves the correspondence of across-nation coding (aggregation to about the "submajor" level in ISCO-88 terms).
- International comparability of ISCO-88 results is improved through technical assistance to participating countries in the use of the system.

## Australia

The Australian Standard Classification of Occupations (ASCO) uses the same concepts of skill level and skill specialization as does ISCO-88 and has eight major groups, 52 minor groups, 282 unit groups, and 1,079 occupations. An occupation is defined as a set of jobs with similar sets of tasks. In 1993, changes likely to be made to ASCO were seen as including (Madden and Tam, 1993):

• Developing procedures for monitoring changes in industry, vocational education, and training to keep ASCO up to date,
• A movement toward the use of competencies (specific skills, knowledge, and training designed to meet industry standards) rather than educational qualifications and duration of training and experience as indicators of skill level,
• Increasing use of job tasks rather than job titles for classifying into occupations, because jobs are becoming broader and titles less reliable indicators of job content, and
• Modifying the major group structure of ASCO to meet user problems, including the need for career path analysis.

### The Netherlands

The Netherlands Standard Classification of Occupations 1992 (NSCO'92) also classifies occupations by skill level and specialization, but it differs primarily in its operational definition of those concepts. This system uses the "most adequate training program, that is the training program that best prepares for the tasks and duties in the job" (Bakker, 1993:273) as the method to identify skill level and specialization for each job. To do this, the Netherlands Standard Classification of Education is used as the basic information to conduct the coding. Skill specialization coding is made according to the major (and minor) educational sectors in the Netherlands, e.g., agriculture, mathematics and natural sciences, and language and culture. Skill level is coded with a five-point scale that combines formal education and length of on-the-job experience. Beyond these higher-level criteria for coding, they include two interesting concepts: main tasks and specific skills. If level and specialization are not adequate, then a list of 128 main tasks is used to differentiate the occupation. Examples of these 128 "task clusters" include "managing supervisors and decision-making general policy," "check, inspect, examine, verify, test, sort," and "navigate a ship." If still further differentiation is needed, then a list of 11 specific skills are used (e.g., quantitative skills defined as activities in which it is important to perform calculations; serviceability defined as activities in which it is important to render service to other people). These concepts are similar to the "generalized work activities" and "basic and cross-functional skills" included in the Occupational Information Network (O*NET™), the Department of Labor's replacement for the *Dictionary of Occupational Titles*. In operation, the five skill levels crossed with the 13 major skill specializations produce 43 occupational "classes."[1] With the invocation of minor skill specializations, 121 occupational groups are formed, and with the addition of the 128 "main tasks" criteria, 1,211 occupations are formed.

## United Kingdom

The United Kingdom replaced two earlier classification systems, the Classification of Occupations and Dictionary of Occupational Titles (CODOT) and the 1980 version of the Classification of Occupations (CO80) with its Standard Occupational Classification. This effort was coincident with the revision of ISCO-68, so an effort was made "to achieve the closest feasible harmonization" between the British SOC and ISCO-88 (White, 1993). Beginning with the 350 entities in the 1980 Classification of Occupations, modifications were made to fit with ISCO classification criteria. These new code groups were tried out against data from the 1981 census of population and a sample of job vacancies sent to job centers. The resulting structure had 9 major groups, 22 submajor groups, 77 minor groups, and 371 occupational unit groups.

## Canada

Canada replaced its Canadian Classification and Dictionary

---

[1] Not 65, as might be expected, since some of the possible cells are not used because of inadequate sample sizes for purposes of statistical reporting. A lower bound of 5,000 job incumbents in the population was set for inclusion of an occupation.

of Occupations (CCDO) in 1991 with "two classifications described as the National Occupational Classification and the Standard Occupational Classification" (Nijhowne and Silver, 1993). The two classifications share a common framework: a hierarchical structure with 514 "unit" groups and 139 "minor" groups in common. The minor groups are organized into 47 major groups in the Standard Occupational Classification (SOC 1991) and into 26 major groups in the National Occupational Classification (NOC 1991). Both have 10 broad occupational categories at the top of the hierarchy.

The SOC 1991 is primarily used for enumeration purposes (e.g., for the Canadian census coding of occupations). The 514 unit groups are described in terms of the principal tasks and duties of the jobs in the unit group. The NOC 1991 also contains other characteristics of the group, such as educational requirements, consistent with its purpose of classifying and describing occupations for labor market transaction. The titles shown as examples are generally the same for the two systems, "but some are unique to each classification" (p. 305). Regarding the military, the SOC 1991 includes just two groups, commissioned officers and other ranks. The NOC 1991 has these two groups, but they include only those military jobs that do not have a civilian counterpart. Military jobs with civilian counterparts are placed in the appropriate occupational unit group within the NOC 1991. The NOC 1991 is used to classify 25,000 job titles into 522 unit groups and does not contain the dictionary-like definitions of its predecessor. Rather, it "serves as a framework whose main function is to provide structure and meaning to the labour market as a whole (Roberts, 1993:320)." One component of NOC 1991 is a matrix, defined by skill level (four levels of type and length of education, training, or experience required for employment in an occupation) and skill type (broadly organized into 9 broad occupational categories, omitting the military category from the 10 broad categories). The 139 minor groups are displayed in this matrix.

## DESCRIPTIVE ANALYTIC SYSTEMS

### Position Analysis Questionnaire

The Position Analysis Questionnaire (PAQ) (McCormick et al., 1969) is a worker-oriented job analysis technique with a long history of research, development, and use with a variety of human resources applications. The PAQ consists of 187 items listing work behaviors and job elements at a level of abstraction that permits work to be described across a broad range of occupations. Completed by subject-matter experts (job incumbents, supervisors, or job analysts who are very familiar with job content), the PAQ reflects a simple model of work performance following an information input—processing—work output sequence. PAQ items are organized into six divisions: information input (e.g., use of written materials), mental processes (e.g., problem solving), work output (e.g., assembling), relationships with other persons (e.g., instructing), job context (e.g., high temperature), and other job characteristics (e.g., work schedule). Five-point response scales are used to assess importance, time spent, extent of use, possibility of occurrence, and applicability of job elements. "Does not apply" is also provided as an option for all items.

The developers of the PAQ provide computerized scoring services and a normative database that permits comparisons of jobs in one organization to similar jobs in other organizations or to all jobs in the database. In addition to summary statistics for job elements (PAQ items), factor analytically derived job dimension scores (e.g., visual input from devices/materials) and estimates of attributes required to perform the job (e.g., visual acuity) can be obtained. This information can be used to estimate the validity of tests of job attributes used for selection purposes. PAQ scores can also be used for job evaluation, to estimate pay rates based on normative wage data for similar jobs in the U.S. economy.

The PAQ has been widely used by human resource professionals and researchers. It has spawned a substantial body of research studies. Major advantages of the instrument include its broad applicability across occupations and availability of the normative database. Numerous criticisms have also been raised by reviewers, including its advanced reading level (a college graduate reading level is required, according to Ash and Edgell, 1975), its content geared too heavily toward manufacturing occupations for an instrument that purports to apply to all jobs (DeNisi et al., 1987), and results that are too general in nature to specify the type of work actually done in a job.

In response to the criticism that the PAQ is too heavily weighted toward blue-collar occupations, in 1986 the PAQ's authors introduced a second worker-oriented instrument called the Professional and Managerial Position Questionnaire (PMPQ). Designed for analysis of managerial, scientific, technical, and staff jobs, the PMPQ consists of 98 items assessing 6 job functions (planning/scheduling, processing of information and ideas, exercising judgment, communicating, interpersonal activities and relationships, and technical activities), personal requirements (e.g., education and training required), and other information (e.g., personnel supervised). As with the PAQ, computerized scoring services and normative data are also available for the PMPQ.

## Fleishman Job Analysis System

The Fleishman Job Analysis System (FJAS) is based on extensive experimental and factor analytic research on the nature of human abilities (Fleishman and Quaintance, 1984). Conducted over a 40-year period, this research program consisted of a wide variety of laboratory tasks designed to elicit performance from subjects drawing on one or more hypothesized underlying abilities. Task batteries were systematically varied to hone in on specific abilities and to delineate the boundaries of their application. Thus, the research linked task characteristics to ability requirements to produce the Fleishman Taxonomy of Human Abilities. The Fleishman taxonomy provides detailed descriptions of 52 abilities, including cognitive (e.g., oral comprehension, number facility), physical (e.g., explosive strength, arm-hand steadiness), psychomotor (e.g., rate control, reaction time), and sensory-perceptual (e.g., depth perception, speech recognition) domains. Nine social-interactive abilities (e.g., persuasion, persistence) and 13 job skills and knowledge (e.g., mechanical knowledge, driving) are the most recent additions (Fleishman, 1992).

A measurement system was also developed to evaluate jobs and tasks for their requisite abilities. The ability requirement scales (Fleishman, 1992) provide definitions, additional information to differentiate each ability from other similar abilities in the taxonomy, and 7-point behaviorally anchored rating scales to aid subject matter experts in estimating the amount of each ability needed to successfully perform a job or task. If tasks are rated, an ability profile for a job can be taken as an average (or weighted average, e.g., by task importance) of abilities required across tasks.

Reliability and interrater agreement are well established (see Fleishman and Mumford, 1988), as is the construct validity of the taxonomy and methods (Fleishman and Mumford, 1989). The FJAS has been especially useful in the development of valid tests linked to job requirements (Fleishman and Mumford, 1988, 1989).

## Occupational Analysis Inventory and the General Work Inventory

The Occupational Analysis Inventory (OAI) is designed to be more relevant to occupational education and guidance, rather than to applied problems in the work setting, which are the focus of systems like the position analysis questionnaire (Cunningham et al., 1983). The inventory includes 617 items, called "work elements," divided across the five categories of information received, mental activities, work behavior, work goals, and work context. Each item is rated on one of four scales: significance, extent, applicability, or a special scale for that element. The three nonspecific scales are relative ratings with adjectivally anchored scale points, e.g., "to a very small extent" at the lower end of the extent scale and "to a great extent" at the higher end. The OAI is characterized as a research tool and it is stated that it is "advisable for the OAI job rater to have college-level reading comprehension, plus some preparatory orientation and practice with the instrument" (Cunningham, 1988:981).

Empirical work has been completed to evaluate the reliability and validity of the OAI (Cunningham, 1988; Cunningham et al., 1983). A study of the reliability of OAI ratings was conducted using 12 job analysts and 21 trained psychology graduate students who rated 215 jobs using written task descriptions from the

U.S. Employment Service. Correlations were computed between two independent raters for each OAI work element. The mean correlation was .53 and the median was .56. Several studies aimed at evaluating the construct validity of the OAI have been conducted, including the comparisons of clusters of occupations obtained with the OAI on several tests and inventories (68 of the 92 measures showed statistically significant discrimination between the clusters), the prediction of mean occupational scores on the General Aptitude Test Battery using OAI factor scores, (median cross-validated multiple correlations were .60 for mental and .24 for motor abilities), bivariate correlations between OAI attribute-requirement estimates and mean scores of job incumbents (statistically significant correlations at the .05 level were found for 38 of 55 analyses), and analyses of variance to relate OAI need-requirement estimates to job satisfaction scores (12 of 15 analyses provided supporting evidence).

The OAI shows generally excellent measurement characteristics, when it is applied in the recommended manner—using college educated, trained analysts. Most of the reported empirical work has been conducted using "paper jobs," that is, written job descriptions from the U.S. Employment Service. It is not clear that it would work as well if used in the field by job incumbents, supervisors, or other occupational experts, many of whom would not be college-trained or be available for special training on the OAI.

A replacement for the OAI, the General Work Inventory (GWI), is shorter and written less technically and could be a more practical alternative for large-scale data collection. This instrument was developed for use by "any literate respondent who is familiar with the job to be analyzed" (Cunningham et al., 1990:34). It has 268 items organized into 8 sections and uses "part of the job" and "extent of occurrence" rating scales, both of which have 9 points and are adjectivally anchored. Research using this inventory in the military showed mean retest reliabilities (for single raters) of .62 across all items, and a mean correlation of profiles of ratings (again, for single raters) of .74, comparable with other similar studies. Ballentine et al. (1992) used the GWI to create a hierarchical structure of Air Force occupations that showed intuitive meaning and corresponded to existing Air Force classifica-

tions, although the comparison was somewhat influenced by ar-
tifactual correspondence between the two systems.

The stream of work represented by the OAI and GWI demonstrates well the utility of using a descriptive system designed to
be applied to the general population of occupations but still retaining enough specificity to provide meaningful differentiations
between occupations, to link to assessments of persons, and to
form useful occupational structures based on the information obtained from the system.

### Common Metric Questionnaire

The common metric questionnaire (CMQ) (The Psychological
Corporation, 1993) was developed by Harvey as a "worker-
oriented" job analysis instrument designed to have applicability
to a broad range of exempt and nonexempt jobs. It is organized
into five major sections (general background, contacts with
people, making decisions, physical and mechanical activities, and
work setting) with several subsections in each. In addition to
general background items that ask about respondent and job characteristics (e.g., tenure in present job, work schedule), the CMQ
consists of 242 behaviorally specific items (e.g., in order to perform your job, do you use desktop or personal computers?). A
matrix response format is used, such that if an item is indicated as
performed, the respondent is asked to provide ratings for up to
four additional scales (e.g., frequency, criticality, consequence of
error). Thus, amount of information provided and amount of
time needed to complete the instrument varies according to job
scope and complexity.

A major advantage of the CMQ, according to its author, is the
possibility of comparing even very dissimilar jobs by virtue of the
instrument's common metric of work descriptors. This may be
useful for purposes of establishing job progression and compensation systems. Broad applicability of the instrument is further
supported by its use of an eighth-grade reading level, so that most
job incumbents can complete it without assistance, and absolute
rather than relative rating scales, so that responses can be compared across jobs. The CMQ can be scored in terms of 80 factor
analytically-derived work dimensions or at the item level, thus

supporting human resource applications requiring relatively abstract (e.g., job classification) or specific (e.g., job descriptions) information.

The CMQ is a recent product, and there does not yet exist a substantial professional literature concerning its usage. The goals that Harvey set for the CMQ, however, particularly concerning ease of use and comparability of data across disparate jobs, are laudable and potentially fill a gap among worker-oriented job analysis instruments that preceded it.

## Multipurpose Occupational Systems Analysis Inventory-Closed Ended

Developed by the Office of Personnel Management, the purpose of the Multipurpose Occupational Systems Analysis Inventory-Closed Ended (MOSAIC) is to collect data on a number of occupational descriptors in a standardized manner across occupations within large occupational families, and then to provide that information in readily accessible electronic databases. MOSAIC has been described as follows: "This system uses an automated occupational analysis approach that eliminates costly redundancies in the collection of data and provides technically sound and legally defensible procedures and documentation to support human resource management (HRM) decisions" (Gregory and Park, 1992:ii).

The report by Gregory and Park illustrates the use of MOSAIC. The occupation focus of the research project was executives, managers, and supervisors. A standard questionnaire was developed and administered to a stratified random, sample of over 20,000 federal executives, managers, and supervisors. The questionnaire contained a diverse set of items, or job descriptors, including: 151 job tasks rated in terms of importance for effective job performance; 22 competencies (a human quality or characteristic associated with the performance of managerial tasks, e.g., knowledge, skill, ability, trait, motive, or self-concept) rated in terms of importance, and needed proficiency at entry; and personal and organizational styles. Data were presented showing the percentage of respondents of various occupational types indicating they performed tasks, found competencies needed at entry, or were important for success. No data on interrater agreement were presented, but a 49 percent return rate was obtained.

## Work Profiling System

The Work Profiling System (WPS), a product of Saville and Holdsworth (1990), is a worker-oriented job analysis instrument supported by expert system computer technology. The WPS is organized into two parts: job tasks and job context. The job tasks section consists of 325 behavior description items (called "tasks") organized into 8 sections (managing tasks, managing people, receiving information, thinking creatively, working with information, communicating, administrating, physical activities) and 30 subsections (e.g., planning/implementing, working with equipment/machinery). Examples of items ("tasks") include: planning a course or route for a journey or voyage; looking after the needs of young children; driving a car, van, or light truck. Items are rated on scales of time spent, importance, and effect of poor performance. Part two, job context, addresses 28 topics, such as education, training, and experience levels needed to perform the job, responsibility for financial resources, types of interpersonal contact, and job-related travel.

Goals for the system include providing an integrated and user-friendly system for job analysis and providing a knowledge base that can serve as the basis for matching people to jobs. Worker attributes are inferred from task ratings using an expert system derived from ratings of attribute-task linkages provided by experienced occupational psychologists. In addition to person-job match, this information base is intended to support such human resource applications as job descriptions, job classification, performance appraisal criteria, job design, and human resource planning.

As is the case with the CMQ, the WPS is a recent product that does not yet have a substantial professional literature concerning its usage. Its objectives as stated by its developers are ambitious, providing a comprehensive methodology for building human resource systems.

# APPENDIX
# C

# Biographical Sketches

**THOMAS A. KOCHAN** *(Co-chair)* is the George M. Bunker professor of management at the Sloan School of Management of the Massachusetts Institute of Technology. He came to MIT from Cornell University, where he was on the faculty of the School of Industrial and Labor Relations from 1973 to 1980. He has served as a third-party mediator, fact finder, and arbitrator and as a consultant to a variety of government and private-sector organizations and labor-management groups. He has done research on a variety of topics related to industrial relations and human resource management in the public and private sector. His recent books include: *After Lean Production: Evolving Employment Practices in the World Auto Industry; Managing for the Future: Organizational Behavior and Processes; Employment Relations in a Changing World Economy; Human Resource Management in Asian Economies; The Mutual Gains Enterprise; Transforming Organizations; An Introduction to Collective Bargaining and Industrial Relations;* and *The Transformation of American Industrial Relations.* In 1988 the *Transformation of American Industrial Relations* received the annual award from the Academy of Management for the best scholarly book on management. He has a Ph.D. in industrial relations from the University of Wisconsin, Madison.

**STEPHEN R. BARLEY** *(Co-Chair)* is professor of industrial engineering and engineering management and the co-director of the Center for Work, Technology and Organization at Stanford University's School of Engineering. Prior to going to Stanford in 1994, Barley served for 10 years on the faculty of the School of Industrial and Labor Relations at Cornell University. He has a Ph.D. in organization studies from the Massachusetts Institute of Technology. He teaches courses on the management of research and development, the organizational implications of technological change, organizational behavior, social network analysis, and ethnographic field methods. He has written extensively on the impact of new technologies on work, the organization of technical work, and organizational culture. He recently edited a volume on technical work, entitled *Between Craft and Science: Technical Work in the United States.* He has served as a consultant to organizations in a variety of industries, including publishing, banking, electronics, and aerospace. He is currently working on a multipronged study of contingent work among engineers and software developers in the Silicon Valley.

**ROSEMARY BATT** is assistant professor of human resource studies at the Industrial and Labor Relations School of Cornell University. She has a B.A. from Cornell University and a Ph.D. from the Sloan School of Management, Massachusetts Institute of Technology. Her research interests include service-sector productivity and competitiveness, strategic human resource management, work design and technology use, and white-collar careers. She has written extensively on the restructuring of the telecommunications and information services industry. She is coauthor of *The New American Workplace: Transforming Work Systems in the United States.*

**NICOLE WOOLSEY BIGGART** is professor of management and sociology at the University of California, Davis. Her research has been concerned largely with the social structure bases of economic organizations. Her book, *Charismatic Capitalism: Direct Selling Organizations in America,* examined the ways in which the direct selling industry makes economic use of the social relations of distributors. She has written about the network relations of the Japanese, South Korean, and Taiwanese economies and is the author (with Gary Hamilton and Marco Orru) of *Economic Organization of East Asia*. Her publications have appeared in the *American Journal of Sociology, Administrative Science Quarterly, Social Problems*, and elsewhere. She is chair of the organizations, occupations, and work section of the American Sociological Association. In 1996 she was the Arthur Andersen distinguished visitor at the Judge Institute of Management Studies, Cambridge University, England. She has a Ph.D. from the University of California, Berkeley.

**PETER CAPPELLI** is professor of management and director of The Wharton School's Center for Human Resources. He has degrees in industrial relations from Cornell University and in labor economics from Oxford, where he was a Fulbright scholar. His research has examined labor relations, changes in work and the effects on skill requirements, the contribution of workplace attitudes and behaviors to job-related skills, and the effects on workforce skills associated with choices of employment practices. His book *Change at Work* describes how the restructuring of American industry has created changes in the employment relationship for the National Planning Association, and *The New Deal at Work* outlines the management challenges that emerge in the absence of long-term employment commitments. He is currently conducting a study of the determinants of financial performance in the life insurance industry and a longitudinal study of the relationship between employment practices and firm performance based on data collection with the U.S. Bureau of the Census.

**MARK J. EITELBERG** is professor of public policy and associate chair for research in the Department of Systems Management at the U.S. Naval Postgraduate School in Monterey, California. He joined the faculty in 1982 after serving as a senior scientist with the Human Resources Research Organization for seven years. He has directed numerous research projects for the Department of Defense and the military services since the mid-1970s and is recognized internationally as a leading scholar in military manpower policy. He is the author or coauthor of more than 100 publications, referenced widely in defense literature. He has served as a consultant and author with several government commissions, defense agencies, and private organizations, including the Brookings Institution, the Atlantic Council of the United States, the RAND Corporation, and the Center for Strategic and International Studies, among others. He is editor of *Armed Forces and Society*, the official journal of the Inter-University Seminar on Armed Forces and Society. He has a Ph.D. in public administration (public policy and national security) from New York University.

**ANN HOWARD** is manager of assessment technology integrity for Development Dimensions International, a leading human resource development firm with over 70 offices and affiliates around the world. Her responsibilities include assuring quality in assessment technologies and providing continuing education for staffing and assessment consultants. Her professional experience includes being president of the Leadership Research Institute, a nonprofit organization that she cofounded in 1987. She was formerly with AT&T, where for 12 years she directed two longitudinal studies of the lives and careers of managers. She has a Ph.D. from the University of Maryland and an M.S. degree from San Francisco State University, both in industrial-organizational psychology. She is the author of more than 75 publications on topics such as assessment centers, management selection, managerial careers, and leadership. She is the senior author (with Douglas W. Bray) of *Managerial Lives in Transition: Advancing Age and Changing Times* and the editor of *The Changing Nature of Work* and *Diagnosis for Organizational Change: Methods and Models*.

**ARNE L. KALLEBERG** is Kenan professor of sociology, and chair of the Department of Sociology at the University of North Carolina, Chapel Hill. He is also an adjunct professor of management in the Kenan-Flagler School of Business at the University of North Carolina and a fellow of the Carolina Population Center. He has a Ph.D. from the University of Wisconsin, Madison. His current research focuses on the changing nature of employment relations in the United States and Norway, organizations' increasing use of flexible staffing arrangements (especially in the health care industry), and the nature and consequences of high-performance work organizations. He has written over 75 articles and chapters, and has coauthored four books, dealing with research topics related to the sociology of work, organizations, occupations and industries, labor markets, and social stratification. His research has been supported by the John Simon Guggenheim Foundation, the Alfred P. Sloan Foundation, the Russell Sage Foundation, the National Science Foundation, the Spencer Foundation, the Japan Foundation, and the Norwegian Research Council.

**ANNE MAVOR** is the study director for the Committee on Techniques for the Enhancement of Human Performance: Occupational Analysis. She is also currently the staff director for the Committee on Human Factors and the Panel on Musculoskeletal Disorders and the Workplace. Her previous work as a National Research Council senior staff officer has included studies on modeling human behavior and command decision making, human factors in air traffic control automation, human factors considerations in tactical display for soldiers, scientific and technological challenges of virtual reality, emerging needs and opportunities for human factors research, and modeling cost and performance for purposes of military enlistment. For the past 25 years, her work has concentrated on human factors, cognitive psychology, and information system design. Prior to joining the National Research Council she worked for the Essex Corporation, a human factors research firm, and served as a consultant to the College Board. She has an M.S. in experimental psychology from Purdue University.

**JAMES P. McGEE** is a senior research associate at the National Research Council. He currently supports the Committee on Techniques for the Enhancement of Human Performance: Occupational Analysis, the Panel on Musculoskeletal Disorders and the Workplace, and the Army Research Laboratory Technical Assessment Board. In addition, he is study director for the Committee on Educational Interventions for Autistic Children. Since 1994 he has supported panels on education and on human factors psychology in such areas as occupational analysis, air traffic control, and military research in human factors. Prior to joining the National Research Council, he held scientific, technical, and management positions in human factors psychology at IBM, RCA, General Electric, General Dynamics, and United Technologies corporations. He has also taught courses in applied psychology at several colleges and is a member of the Potomac Chapter of the Human Factors and Ergonomics Society. He has a Ph.D. in experimental psychology from Fordham University.

**DAVID NEUMARK** is professor of economics at Michigan State University and a research associate of the National Bureau of Economic Research. He has a bachelors degree in economics from the University of Pennsylvania and M.A. and Ph.D. degrees from Harvard University, with a specialization in labor economics and econometrics. He worked at the Federal Reserve Board and the University of Pennsylvania before coming to Michigan State in 1994. His current research interests cover job stability, minimum wages, affirmative action, discrimination, aging, and supplemental security income.

**PAUL OSTERMAN** is professor of human resources and management at the Sloan School of Management of the Massachusetts Institute of Technology. He is the author of three books: *Getting Started: The Youth Labor Market, Employment Futures: Reorganization, Dislocation, and Public Policy,* and *Making America Work*; the coauthor of *The Mutual Gains Enterprise, Forging a Winning Partnership Among Labor, Management, and Government,* and the editor of two books, *Internal Labor Markets* and *Broken Ladders: Managerial Careers in the New Economy*. In addition, he has written numerous academic journal articles and policy issue papers on such topics as labor market policy, job training programs, economic development, antipoverty programs, and the organization of work within firms. He has been a senior administrator of job training programs for the Commonwealth of Massachusetts and consulted widely for government agencies, foundations, community groups, and public interest organizations. He has a Ph.D. in economics from MIT.

**NORMAN G. PETERSON** is senior research fellow at the American Institutes for Research (AIR). He has B.A. and Ph.D. degrees from the University of Minnesota in psychology, specializing in industrial and organizational psychology. Prior to joining AIR, he was vice-president of the Personnel Decisions Research Institute. He is a fellow of the Society for Industrial and Organizational Psychology, the American Psychological Association, and the American Psychological Society. He has conducted research for a variety of public- and private-sector sponsors in a number of applied areas, including occupational analysis, development and validation of measures of individual differences, employee selection and classification systems, and the prediction of human performance in occupational and training settings.

**LYMAN W. PORTER** is professor of management in the Graduate School of Management at the University of California, Irvine, and was formerly dean of that school. Previously he served on the faculty of the University of California, Berkeley. Currently, he serves as a member of the academic advisory board of the Czechoslovak Management Center, a member of the board of trustees of the American University of Armenia, and was formerly an external examiner for the National University of Singapore. He is a past president of the Academy of Management and has received both its Scholarly Contributions to Management Award and its Distinguished Management Educator Award. He has also served as president of the Society of Industrial-Organizational Psychology, and in 1989 received its Distinguished Scientific Contributions Award. His major fields of interest are organizational psychology, management, and management education and has written extensively in these fields. His 1988 book (with Lawrence McKibbin), *Management Education and Development*, reported the findings of a nationwide study of business school education and post-degree management development.

**KENNETH I. SPENNER** is professor and chair of the Department of Sociology at Duke University. He has a bachelors degree from Creighton University, a masters degree from the University of Notre Dame, and a Ph.D. in sociology from the University of Wisconsin, Madison. His research interests include work and personality, occupations, career dynamics, technology, and the sociology of organizations and markets. His major current research project is a multiyear panel study of organizational adaptation and survival of a large sample of formerly state-owned manufacturing enterprises in Bulgaria.

**LTG THEODORE G. STROUP, JR.**, is the vice president, education and executive director of the Institute of Land Warfare for the Association of the U.S. Army, a nonprofit educational association in Arlington, Virginia. General Stroup served for 34 years in the U.S. Army in multiple worldwide assignments in peace and war. His primary military specialty was as a combat engineer; his last 10 years of service as a flag officer were primarily in strategic resource planning, manpower, and human resources. He was at retirement the U.S. Army deputy chief of staff of personnel. He has a B.S. from the United States Military Academy at West Point, an M.S.E. from Texas A&M, and an M.B.A. from the American University in the fields of economics and finance. He is a licensed professional engineer and has authored articles on military manpower, training, leadership, and military engineering.

**ROBERT J. VANCE** is senior research associate at the Institute for Policy Research and Evaluation at the Pennsylvania State University. He has a B.A. in psychology from the University of Connecticut and M.S. and Ph.D. degrees in industrial and organizational psychology from the Pennsylvania State University. Prior to joining Penn State in 1990, he served on the psychology faculty at Ohio State University. His research and teaching interests are in the areas of personnel selection, job performance measurement, work motivation, and organizational development.

# Index

O*NET™, 185-186, 279, 280, 324
*see also* Demographic factors;
    Economic factors; Market
    forces; Organizational
    factors; Technological
    factors
Ethnography, *see* Anthropological
    approach

## F

Fair Labor Standards Act, 108
Families of jobs, *see* Job families
Family and Medical Leave Act, 18
Family factors, 8, 16, 17, 18, 28
    family-operated enterprises, 179
    military personnel, 220, 225, 281
    networking, 89-90
Farm workers, *see* Agricultural
    workers
Federal government
    occupational classification/
        analysis systems, not SOC,
        167; *see also* Army Personnel
        Network (AP*NET);
        Occupational Information
        Network (O*NET™)
    SOC, 7, 168-169, 174, 176, 177-181,
        333
    *see also specific departments and
        agencies; terms beginning
        "Military..."*
Fertility, 43
Financial markets, 33-35, 267
Fleishman Job Analysis System, 172,
    173, 186, 335-336
*Force XXI Operations*, 251, 252, 255-256
Foreign countries, *see* International
    perspectives

## G

Gender factors, 3
    age factors and, 43, 44, 225
    attitudes toward work, 50-51

education and training, 96
emotional work, 201
historical perspectives, 16, 28, 41-
    49 (passim), 63-66, 70
military personnel, 8, 223, 224-225,
    227-230, 281
nonstandard employment, 91
occupational classification
    systems, 61-62
service workers, 121, 133
unpaid work, 26
workforce, general, 16, 28, 41-49
    (passim), 63-66, 70
worklife expectancy, 45-47
work timing, 57
General Accounting Office, 85
General Aptitude Text Battery, 327
General Social Survey, 49-51, 56-57
General Work Inventory, 172, 173,
    337-338
Germany, 39
Globalization of markets, 16-17, 31,
    71, 266-267
    blue-collar workers, 117-119
    compensation, 31, 32, 266
    downsizing and, 17, 32
    financial markets, 34, 35-36
    foreign direct investment, 32, 118
    organizational factors, 32, 35
    technological factors, 31, 32-33
Group performance, *see* Team skills;
    Teamwork

## H

*Handbook for Analyzing Jobs*, 182
High-performance work systems, 109-
    110, 130, 189, 264, 272, 274,
    278
Hiring, *see* Recruitment, military;
    Selection and placement
Hispanics, 43, 45, 64-66
    military personnel, 223, 230-232
Hours of work, 4, 47-49, 56, 58, 275
    part-time employment, 57-58, 92, 93

# DATE DUE

|  |  |  |  |
|---|---|---|---|
|  |  |  |  |
|  |  |  |  |
|  |  |  |  |
|  |  |  |  |
|  |  |  |  |
|  |  |  |  |
|  |  |  |  |
|  |  |  |  |
|  |  |  |  |
|  |  |  |  |
|  |  |  |  |
|  |  |  |  |
|  |  |  |  |
|  |  |  |  |
|  |  |  |  |
|  |  |  |  |
|  |  |  |  |
|  |  |  |  |
|  |  |  |  |
|  |  |  |  |
| GAYLORD |  |  | PRINTED IN U.S.A. |